Created by Xspurts.com

All rights reserved.

Copyright © 2005 onwards .

By reading this book, you agree to the below Terms and Conditions.

Xspurts.com retains all rights to these products.

No part of this book may be reproduced in any form, by photostat, microfilm, xerography, or any other means, or incorporated into any information retrieval system, electronic or mechanical, without the written permission of Xspurts.com; exceptions are made for brief excerpts used in published reviews.

This publication is designed to provide accurate and authoritative information with regard to the subject matter covered but is for entertainment purposes only. It is sold with the understanding that the publisher is not engaged in rendering legal, accounting, health, relationship or other professional / personal advice. If legal advice or other expert assistance is required, the services of a competent professional should be sought.

♥ A New Zealand Designed Product

Get A Free Book At: xspurts.com/posts/free-book-offer

Table of Contents

Table of Contents
Understanding The Concept Of Natural Order
Definition of Natural Order
Historical Overview of Natural Order
Philosophical Insights Into Natural Order
Antiquity Period and the Natural Order
Medieval Philosophy and the Natural Order
The Enlightenment Era's Perspective on Natural Order
Political Philosophy and the Natural Order
Scientific Developments and the Natural Order
Natural Order in Eastern Philosophy
Natural Order in Confucianism
Taoism's View on the Natural Order
Natural Order and Theism
Natural Order in Monotheist Traditions
Pantheism and the Natural Order
Evolutionary Theories and Natural Order
The Darwinian Principle and its Philosophy
Natural Order and Genetics
The Intersection of Natural Order and Ethics
Ethical Decisions in Alignment with the Natural Order
Relativism versus Constant Ethics in Natural Order
Natural Order as Depicted in Literature
Shakespeare's Contribution to the Understanding of Natural Order
Modern Literary Takes on the Natural Order
Natural Order in Political Philosophy
Political Systems and the Natural Order
Anarchism and the Philosophy of Natural Order
Natural Order and Environmental Ethics

Environmental Balance and Natural Order
Human Impacts on the Natural Order
Natural Order and Quantum Physics
Reality and the Perception of Natural Order
The Role of Uncertainty in Natural Order
Contemporary Criticisms of the Natural Order Concept
The Naturalistic Fallacy
Postmodernist Views
Natural Order and Artificial Intelligence
Technology and the Challenge of Natural Order
Ethical Questions in the Age of AI
Reconceptualizing Natural Order in the 21st Century
Sustainability and the Natural Order
Prospects for the Future of Natural Order
The Future of Natural Order Philosophy
The Role of the Natural Order in Future Ethics
Natural Order in the Age of Climate Change
Have Questions / Comments?
Get Another Book Free

Understanding The Concept Of Natural Order

Natural order, in the realm of philosophy, encapsulates the underlying principles and patterns that govern the universe. It's a concept deeply ingrained in various philosophical traditions, ranging from ancient Greek philosophy to modern-day existentialism. At its core, natural order seeks to elucidate the harmony and coherence present in the natural world, offering insights into the structure and functioning of reality.

One of the fundamental aspects of natural order is the idea of balance and equilibrium. This notion suggests that the universe operates according to certain laws and principles that maintain a state of equilibrium. For instance, in ancient Greek philosophy, the concept of "cosmos" referred to the ordered and harmonious arrangement of the universe, governed by divine laws. Similarly, Eastern philosophies such as Taoism emphasize the importance of achieving balance between opposing forces, known as yin and yang, to maintain harmony in the cosmos.

Another key element of natural order is the concept of causality. According to this principle, every event in the universe is interconnected and influenced by preceding events, creating a chain of cause and effect. This idea has profound implications for understanding the unfolding of events in the natural world and has been explored extensively in philosophical discourse. For instance, in the Western philosophical tradition, thinkers like Aristotle emphasized the principle of causality as a fundamental aspect of natural order, shaping our understanding of the physical and metaphysical realms.

Furthermore, natural order encompasses the notion of hierarchy and structure within the universe. This hierarchical arrangement delineates the relationships between different entities, with each occupying a distinct position within the cosmic order. In ancient Indian philosophy, the concept of "dharma" encapsulates the moral and cosmic order that governs the universe, establishing a framework for ethical conduct and social harmony. Similarly, in Western philosophy, thinkers like Plato and Aristotle elucidated hierarchical systems of being, wherein entities are classified based on their intrinsic qualities and capacities.

Moreover, natural order encompasses the idea of teleology, which posits that there is an inherent purpose or end goal guiding the unfolding of events in the universe. This teleological perspective suggests that phenomena in nature are directed towards certain outcomes, imbuing the cosmos with a sense of purpose and direction. This concept has been explored in various philosophical traditions, with thinkers like Immanuel Kant proposing teleological arguments to elucidate the order and purpose inherent in the natural world.

In contemporary times, the concept of natural order continues to be a subject of philosophical inquiry, albeit with new insights and perspectives. From existentialist thinkers questioning the existence of inherent meaning in the universe to ecological philosophers advocating for a deeper understanding of humanity's interconnectedness with nature, the discourse surrounding natural order remains vibrant and multifaceted.

In conclusion, natural order represents a multifaceted concept that underpins our understanding of the universe. From notions of balance and causality to ideas of hierarchy and teleology, the concept of natural order offers profound insights into the structure and functioning of reality. Across different philosophical traditions and epochs, the pursuit of understanding natural order has been a central endeavor, shaping our perceptions of the cosmos and our place within it.

Definition of Natural Order

Natural order, within the domain of philosophy, denotes a concept deeply entrenched in the exploration of the fundamental principles governing the universe. It encompasses the inherent patterns, structures, and laws that regulate the natural world, offering insight into the organization and functioning of reality.

At its core, the definition of natural order revolves around the idea of coherence and harmony within the cosmos. It posits that there exists an intrinsic orderliness in the universe, characterized by a systematic arrangement of elements and processes. This notion suggests that the phenomena observed in nature are not random but instead follow discernible patterns and principles.

One of the key aspects of the definition of natural order is the principle of causality. According to this principle, every event is causally connected to preceding events, forming a chain of cause and effect. This causal relationship underscores the interconnectedness of phenomena in the natural world, highlighting the deterministic nature of reality. From the falling of a leaf to the orbit of celestial bodies, causality permeates all aspects of existence, shaping the unfolding of events in the universe.

Furthermore, the definition of natural order encompasses the concept of balance and equilibrium. It suggests that the universe operates in a state of equilibrium, wherein opposing forces are balanced and harmonized. This balance is evident in various aspects of nature, such as the delicate ecosystems of our planet or the intricate balance of forces in the cosmos. Maintaining this equilibrium is crucial for the sustainability and stability of the natural world.

Another facet of the definition of natural order is the idea of hierarchy and structure. It postulates that there exists a hierarchical organization within the universe, wherein entities are classified based on their intrinsic qualities and relationships. This hierarchical structure delineates the order of existence, from subatomic particles to galaxies, each occupying a specific position within the cosmic hierarchy.

Moreover, the definition of natural order encompasses the concept of teleology, which pertains to the inherent purpose or end goal guiding the phenomena in the universe. This teleological perspective suggests that events in nature are directed towards certain outcomes or ends, imbuing the cosmos with a sense of purpose and direction. This notion of purposefulness underscores the idea that there is an underlying intentionality driving the unfolding of events in the natural world.

In contemporary philosophical discourse, the definition of natural order continues to be a subject of inquiry and debate. With advancements in various fields such as physics, biology, and ecology, our understanding of the principles governing the universe has evolved significantly. From chaos theory to complexity science, new paradigms have emerged that challenge traditional conceptions of natural order, prompting a reevaluation of our understanding of reality.

In conclusion, the definition of natural order encompasses a multifaceted concept that elucidates the underlying principles governing the universe. From causality and balance to hierarchy and teleology, natural order offers profound insights into the organization and functioning of reality. As our understanding of the natural world continues to evolve, so too will our comprehension of the intricate and interconnected systems that define the cosmos.

Historical Overview of Natural Order

Throughout human history, the concept of natural order has been a central theme in philosophical inquiry, shaping our understanding of the universe and our place within it. From ancient civilizations to modern thinkers, the notion of a fundamental order governing the cosmos has been a recurring motif, influencing various philosophical, religious, and scientific traditions.

Ancient civilizations, such as those of Mesopotamia, Egypt, and Greece, developed early notions of natural order rooted in mythological and religious beliefs. In Mesopotamia, the Babylonians conceived of the cosmos as a structured hierarchy, with gods presiding over different aspects of nature. Similarly, ancient Egyptian mythology depicted a world governed by divine principles, where cosmic order was maintained through rituals and offerings to the gods. In Greece, philosophers like Heraclitus and Pythagoras pondered the concept of a harmonious cosmos governed by mathematical principles, laying the groundwork for later philosophical developments.

The ancient Greeks made significant contributions to the concept of natural order through their philosophical inquiries. Plato, in his dialogue "Timaeus," proposed the idea of a demiurge—a divine craftsman who imposed order and structure upon the chaotic primordial matter, creating the cosmos according to rational principles. Aristotle, in his "Physics" and "Metaphysics," further developed the notion of natural order, emphasizing the teleological aspect of nature and the inherent purposefulness of natural phenomena.

During the Middle Ages, the concept of natural order underwent significant transformations as it became intertwined with Christian theology. Christian thinkers, such as Thomas Aquinas, sought to reconcile the teachings of Christianity with the philosophical insights of Aristotle, thereby integrating the notion of natural order into Christian doctrine. According to Aquinas, God imbued the universe with order and purpose, which could be discerned through reason and revelation.

The Renaissance period witnessed a resurgence of interest in natural philosophy and the exploration of the natural world. Thinkers like Galileo Galilei and Johannes Kepler revolutionized our understanding of the cosmos through their observations and mathematical analyses, laying the groundwork for modern science. Galileo's heliocentric model of the solar system and Kepler's laws of planetary motion provided compelling evidence for the existence of a natural order governing the movements of celestial bodies.

The Enlightenment era ushered in a new era of scientific inquiry and philosophical speculation, challenging traditional conceptions of natural order. Thinkers like Isaac Newton and René Descartes formulated comprehensive theories of the universe based on mathematical principles and mechanistic explanations. Newton's laws of motion and universal gravitation revolutionized our understanding of the physical world, demonstrating the existence of a unified natural order governing both terrestrial and celestial phenomena.

In the modern era, advances in fields such as physics, biology, and ecology have deepened our understanding of natural order. The development of quantum mechanics and chaos theory has revealed the intricate and dynamic nature of the universe, challenging deterministic conceptions of natural order and highlighting the role of randomness and complexity in shaping natural phenomena.

In conclusion, the historical overview of natural order traces the evolution of this concept from ancient mythological beliefs to modern scientific theories. Throughout history, the notion of a fundamental order governing the cosmos has been a source of fascination and inquiry, shaping the development of philosophy, religion, and science. As our understanding of the natural world continues to evolve, so too will our appreciation of the intricate and interconnected systems that define the universe.

Philosophical Insights Into Natural Order

Throughout the history of philosophy, the concept of natural order has been a subject of profound inquiry and contemplation, leading thinkers to explore the underlying principles that govern the universe and guide human existence. From ancient philosophers to contemporary thinkers, various insights into natural order have emerged, shaping philosophical discourse and influencing our understanding of the world.

One of the earliest philosophical insights into natural order can be found in the works of ancient Greek philosophers such as Heraclitus and Pythagoras. Heraclitus, known for his doctrine of flux, posited that the universe is in a constant state of change, yet governed by an underlying order or "logos." According to Heraclitus, this logos, or rational principle, gives coherence to the ever-changing flux of reality, providing a framework for understanding the unity and harmony of the cosmos. Pythagoras, on the other hand, saw natural order as inherently mathematical, believing that the universe is structured according to numerical ratios and proportions, which govern the relationships between objects and phenomena.

Plato, in his dialogues, offered profound insights into the nature of natural order, particularly in his metaphysical works such as the "Timaeus." Plato conceived of the universe as a harmonious and intelligible cosmos, created by a divine craftsman or demiurge who imposed order upon the pre-existing chaos. According to Plato, the universe is governed by eternal and immutable forms, which serve as the blueprint for all existence, providing a transcendent basis for natural order and intelligibility.

Aristotle, in his philosophical treatises, further developed the concept of natural order, emphasizing the teleological aspect of nature and the inherent purposefulness of natural phenomena. In Aristotle's view, everything in the natural world has a final cause or end towards which it tends, reflecting the overarching order and harmony of the cosmos. Aristotle's teleological understanding of natural order influenced subsequent philosophical and theological traditions, shaping the medieval concept of the Great Chain of Being and the Christian notion of divine providence.

In the modern era, thinkers such as René Descartes and Immanuel Kant offered new perspectives on natural order, grounded in the principles of rationalism and empiricism. Descartes, in his "Meditations on First Philosophy," sought to establish a secure foundation for knowledge based on clear and distinct ideas, thereby grounding natural order in the rational structure of the mind. Kant, in his "Critique of Pure Reason," argued that natural order is not inherent in the world itself but imposed by the human mind through the categories of understanding, which structure our experience of reality.

Contemporary philosophers continue to explore the concept of natural order from various perspectives, drawing insights from fields such as complexity theory, systems theory, and environmental philosophy. The emergence of interdisciplinary approaches to natural order reflects a growing recognition of the interconnectedness and complexity of natural systems, challenging traditional conceptions of linear causality and determinism.

In conclusion, philosophical insights into natural order have evolved over millennia, reflecting humanity's ongoing quest to understand the fundamental principles that govern the universe. From ancient Greece to the present day, thinkers have grappled with questions about the nature of reality, the structure of the cosmos, and the role of human beings within the larger scheme of things. While philosophical perspectives on natural order may vary, the quest for knowledge and understanding remains a central driving force in philosophical inquiry.

Antiquity Period and the Natural Order

The Antiquity Period, spanning from around the 8th century BCE to the 5th century CE, was a pivotal era in the development of philosophical thought and the exploration of the concept of natural order. During this time, philosophers in ancient Greece and Rome grappled with fundamental questions about the nature of reality, the order of the cosmos, and the role of human beings within the natural world.

One of the most influential figures of this period was Heraclitus of Ephesus, known for his doctrine of flux and the concept of the "logos." Heraclitus proposed that everything is in a constant state of change, and that underlying this flux is a universal principle of order or "logos." According to Heraclitus, the logos provides coherence and unity to the ever-changing world, establishing a fundamental harmony within the natural order.

Pythagoras, a contemporary of Heraclitus, offered a different perspective on natural order, emphasizing the mathematical principles that govern the universe. Pythagoras believed that the cosmos is structured according to numerical ratios and proportions, which underlie the harmony and order of the natural world. This mathematical understanding of natural order profoundly influenced later thinkers, shaping the development of mathematics and scientific inquiry.

Plato, in his philosophical dialogues, further elaborated on the concept of natural order, particularly in works such as the "Timaeus." Plato conceived of the universe as a rational and intelligible cosmos, created by a divine craftsman or demiurge who imposed order upon the pre-existing chaos. According to Plato, the universe is governed by eternal and immutable forms, which serve as the blueprint for all existence, providing a transcendent basis for natural order and intelligibility.

Aristotle, a student of Plato, made significant contributions to the understanding of natural order in his philosophical treatises. Aristotle emphasized the teleological aspect of nature, arguing that everything in the natural world has a final cause or end towards which it tends. In Aristotle's view, natural order is inherently purposeful, reflecting the overarching harmony and design of the cosmos. His teleological understanding of natural order laid the groundwork for subsequent philosophical and theological traditions.

The Roman period also witnessed significant developments in the exploration of natural order, particularly through the works of philosophers such as Lucretius and Seneca.

Lucretius, in his epic poem "De Rerum Natura" (On the Nature of Things), presented a materialistic worldview grounded in the principles of atomism and determinism. According to Lucretius, the universe is composed of indivisible atoms that interact according to natural laws, resulting in the emergence of order and complexity.

Seneca, a Stoic philosopher, explored the concept of natural order in his ethical writings, emphasizing the importance of living in accordance with nature and accepting the inevitable changes of the world. For Seneca, natural order is synonymous with divine providence, guiding human beings towards wisdom and virtue.

In conclusion, the Antiquity Period was a rich and fertile period in the exploration of natural order, with philosophers in ancient Greece and Rome offering diverse perspectives on the fundamental principles that govern the universe. From Heraclitus' doctrine of flux to Plato's concept of eternal forms, and from Aristotle's teleological understanding to the materialistic worldview of Lucretius, the thinkers of antiquity laid the foundation for centuries of philosophical inquiry into the nature of reality and the order of the cosmos.

Medieval Philosophy and the Natural Order

The Medieval Period, spanning roughly from the 5th century CE to the 15th century CE, was characterized by profound intellectual and philosophical developments that significantly influenced the understanding of the natural order. During this time, philosophers and theologians in Europe grappled with questions about the relationship between God, nature, and humanity, drawing on both classical philosophical traditions and Christian theology to explore the concept of natural order.

One of the most influential figures of Medieval philosophy was Saint Augustine of Hippo, whose theological writings had a lasting impact on Western thought. Augustine's conception of natural order was deeply influenced by his Christian faith, as well as by his engagement with classical philosophy, particularly Neoplatonism. Augustine argued that the natural order is governed by divine providence, with God as the ultimate source of order, goodness, and harmony in the universe. According to Augustine, the created world reflects the wisdom and goodness of its Creator, and human beings are called to live in harmony with the natural order as part of their journey towards God.

Another key figure in Medieval philosophy was Thomas Aquinas, a Dominican friar and theologian known for his synthesis of Christian theology and Aristotelian philosophy. Aquinas's concept of natural order was grounded in his understanding of the relationship between reason and faith, as well as his belief in the existence of a rational and intelligible cosmos created by God. According to Aquinas, the natural order is characterized by a hierarchy of being, with each level of existence reflecting the divine plan and purpose. Human reason, guided by divine revelation, allows individuals to discern the natural order and live virtuously in accordance with God's will.

Medieval philosophers also explored the concept of natural law, which played a central role in ethical and political thought during this period. Natural law, derived from the rational order of the universe, was seen as the basis for moral principles that are inherent in human nature and accessible to reason. Figures such as Duns Scotus and William of Ockham further developed the idea of natural law, emphasizing the importance of individual freedom and conscience in moral decision-making.

The Medieval Period also saw the emergence of new scientific and philosophical perspectives that challenged traditional understandings of the natural order. Figures such as Roger Bacon and Albertus Magnus made significant contributions to the study of natural philosophy, advocating for empirical observation and experimentation as means of understanding the natural world. These early scientific endeavors laid the groundwork for the Scientific Revolution that would unfold in the subsequent centuries.

In summary, Medieval philosophy was a dynamic and diverse intellectual movement that shaped the Western understanding of the natural order. From Augustine's theological reflections to Aquinas's synthesis of faith and reason, and from the development of natural law theory to the early scientific inquiries of Bacon and Magnus, Medieval thinkers grappled with complex questions about the relationship between God, nature, and humanity, leaving a lasting legacy that continues to influence philosophical and theological discourse to this day.

The Enlightenment Era's Perspective on Natural Order

The Enlightenment Era, also known as the Age of Reason, marked a significant shift in philosophical thought, challenging traditional beliefs and ushering in new perspectives on the concept of natural order. Lasting roughly from the late 17th century to the early 19th century, this period was characterized by an emphasis on reason, empiricism, and skepticism, as well as a growing interest in science, politics, and human rights.

One of the central themes of Enlightenment philosophy was the rejection of supernatural explanations in favor of rational inquiry and empirical observation. Enlightenment thinkers sought to understand the natural world through the application of reason and scientific method, leading to groundbreaking discoveries in fields such as physics, astronomy, and biology. Figures such as Isaac Newton, Galileo Galilei, and Rene Descartes laid the foundations for modern science, challenging traditional religious and metaphysical conceptions of the natural order.

Enlightenment philosophers also developed new theories about the origins and organization of the universe, challenging prevailing religious and theological explanations. The French philosopher and mathematician, Pierre-Simon Laplace, famously proposed the nebular hypothesis, which suggested that the solar system formed from a rotating disk of gas and dust—a radical departure from traditional creation myths.

In addition to scientific advancements, Enlightenment thinkers also explored the concept of natural order in the realm of politics and society. Influenced by the social contract theory of thinkers such as Thomas Hobbes and John Locke, Enlightenment philosophers argued that human societies are governed by natural laws that can be understood through reason and observation. These natural laws, they contended, dictate the rights and responsibilities of individuals and the proper organization of political institutions.

One of the most influential Enlightenment philosophers was Immanuel Kant, whose work on epistemology and ethics had a profound impact on modern philosophy. Kant proposed that the natural order is not merely a product of empirical observation but is also shaped by the inherent structure of human cognition. According to Kant, the mind imposes certain categories and concepts onto sensory experience, giving rise to our understanding of the natural world.

The Enlightenment Era also saw the emergence of new philosophical movements, such as utilitarianism and existentialism, which further explored the concept of natural order from different perspectives. Utilitarian philosophers like Jeremy Bentham and John Stuart Mill argued that the natural order is governed by the principle of utility, which seeks to maximize happiness and minimize suffering. Existentialist thinkers like Jean-Paul Sartre and Friedrich Nietzsche, on the other hand, challenged traditional notions of natural order, emphasizing the individual's freedom to create meaning and purpose in a seemingly chaotic and indifferent universe.

In conclusion, the Enlightenment Era brought about a fundamental reevaluation of the concept of natural order, challenging traditional religious and metaphysical explanations in favor of rational inquiry and empirical observation. Through advances in science, philosophy, and politics, Enlightenment thinkers laid the groundwork for modern understandings of the natural world and human society, shaping the course of intellectual history for centuries to come.

Political Philosophy and the Natural Order

Political philosophy and the concept of the natural order have been intertwined throughout history, influencing the way societies are organized and governed. From ancient times to the present day, political thinkers have grappled with questions about the nature of political authority, the rights and duties of citizens, and the proper role of government within the natural order.

In ancient Greece, political philosophers such as Plato and Aristotle explored the relationship between natural order and political organization. Plato's Republic envisioned a utopian society governed by philosopher-kings, where each individual's role in society was determined by their natural abilities and inclinations. Aristotle, meanwhile, emphasized the importance of the natural order in his Politics, arguing that the best form of government is one that reflects the natural hierarchy of society.

During the Renaissance, political philosophers like Niccolò Machiavelli and Thomas Hobbes challenged traditional conceptions of political authority and the natural order. Machiavelli's seminal work, The Prince, advocated for a pragmatic approach to politics, arguing that rulers should prioritize power and stability over moral considerations. Hobbes, in his famous work Leviathan, proposed a social contract theory in which individuals voluntarily surrender certain freedoms to a sovereign authority in exchange for protection and security.

The Enlightenment Era saw further developments in political philosophy and the concept of the natural order. Enlightenment thinkers such as John Locke and Jean-Jacques Rousseau advanced theories of natural rights and popular sovereignty, laying the groundwork for modern liberal democracy. Locke's Second Treatise of Government argued that individuals possess natural rights to life, liberty, and property, which governments are duty-bound to protect. Rousseau, in his Social Contract, proposed a vision of direct democracy based on the general will of the people.

In the 19th and 20th centuries, political philosophers continued to explore the relationship between natural order and political organization in response to the challenges posed by industrialization, urbanization, and globalization. Karl Marx, for example, critiqued capitalist society as a distortion of the natural order, arguing that the inherent contradictions of capitalism would ultimately lead to its downfall. His vision of a classless society governed by the principle of "from each according to his ability, to each according to his needs" remains influential to this day.

In the 20th century, political philosophers such as John Rawls and Robert Nozick reinvigorated debates about justice, equality, and the natural order. Rawls's A Theory of Justice proposed a theory of justice based on the principles of fairness and equality, while Nozick's Anarchy, State, and Utopia defended a minimalist state based on individual rights and voluntary exchange.

Today, political philosophers continue to grapple with questions about the natural order and its implications for political organization and governance. Debates about issues such as distributive justice, environmental ethics, and the role of technology in society reflect ongoing concerns about how to balance individual freedoms with collective well-being within the natural order.

In conclusion, political philosophy and the concept of the natural order have long been intertwined, shaping the way societies are organized and governed. From ancient Greece to the present day, political thinkers have explored questions about the nature of political authority, the rights and duties of citizens, and the proper role of government within the natural order, leaving a rich legacy of ideas that continue to inform contemporary debates about politics and society.

Scientific Developments and the Natural Order

Scientific developments have long played a significant role in shaping our understanding of the natural order, influencing not only how we perceive the world around us but also our place within it. From ancient observations of celestial bodies to modern discoveries in quantum mechanics, science has provided valuable insights into the fundamental laws and patterns that govern the universe.

One of the earliest scientific contributions to the understanding of the natural order came from the ancient Greeks. Philosophers such as Pythagoras and Aristotle sought to explain natural phenomena through rational inquiry and observation. Pythagoras, for example, proposed that the universe was governed by mathematical principles, laying the foundation for later scientific inquiry into the natural order.

During the Renaissance, the Scientific Revolution ushered in a new era of discovery and exploration. Pioneering scientists such as Nicolaus Copernicus, Johannes Kepler, and Galileo Galilei challenged prevailing beliefs about the natural order, leading to groundbreaking discoveries in astronomy and physics. Copernicus's heliocentric model of the solar system, Kepler's laws of planetary motion, and Galileo's telescopic observations of the heavens revolutionized our understanding of the cosmos and our place within it.

In the 17th and 18th centuries, the Enlightenment further advanced scientific inquiry and understanding of the natural order. Thinkers such as Isaac Newton and Rene Descartes laid the groundwork for modern physics and mathematics, developing theories and methodologies that continue to shape scientific inquiry today. Newton's laws of motion and universal gravitation, in particular, provided a comprehensive framework for understanding the natural world and its underlying order.

The 19th century saw significant advancements in biology, chemistry, and geology, further illuminating the intricacies of the natural order. Charles Darwin's theory of evolution by natural selection revolutionized our understanding of life on Earth, demonstrating how species evolve and adapt to their environments over time. In chemistry, Dmitri Mendeleev's periodic table of elements provided a systematic classification of the building blocks of matter, revealing underlying patterns and relationships within the natural world.

In the 20th and 21st centuries, scientific developments have accelerated at an unprecedented pace, leading to new discoveries and insights into the natural order. Quantum mechanics, relativity theory, and genetics have revolutionized our understanding of the fundamental laws of physics, challenging conventional notions of causality and determinism. The advent of technologies such as the electron microscope, the Hubble Space Telescope, and the Large Hadron Collider has enabled scientists to explore the universe on both the smallest and largest scales, unveiling hidden dimensions of the natural order.

Moreover, interdisciplinary research has become increasingly important in unraveling the complexities of the natural order. Fields such as systems biology, complexity theory, and network science seek to understand how complex systems emerge and evolve across different scales of organization, from the molecular to the ecological.

In conclusion, scientific developments have played a crucial role in shaping our understanding of the natural order, from ancient times to the present day. Through observation, experimentation, and theoretical inquiry, scientists have uncovered fundamental laws and patterns that govern the universe, shedding light on our place within it and inspiring further exploration and discovery. As we continue to push the boundaries of scientific knowledge, the study of the natural order remains an essential endeavor, offering valuable insights into the mysteries of existence.

Natural Order in Eastern Philosophy

Eastern philosophy, encompassing traditions such as Hinduism, Buddhism, Taoism, and Confucianism, offers profound insights into the concept of the natural order. Unlike Western philosophical traditions, which often focus on rational inquiry and empirical observation, Eastern philosophy places greater emphasis on intuition, contemplation, and harmony with nature. Within these traditions, the natural order is understood as an interconnected web of relationships and patterns that underlie all existence.

In Hinduism, the concept of dharma plays a central role in understanding the natural order. Dharma refers to the moral and ethical duties that govern individual conduct and societal harmony. According to Hindu philosophy, all beings are part of a cosmic order (rita) that sustains the universe and maintains balance. Dharma encompasses not only ethical behavior but also one's cosmic duty (svadharma) based on one's caste, stage of life, and spiritual path. By adhering to dharma, individuals contribute to the maintenance of the natural order and promote the welfare of all beings.

Similarly, in Buddhism, the natural order is viewed through the lens of dependent origination (pratītyasamutpāda), which teaches that all phenomena arise interdependently and are impermanent. According to Buddhist philosophy, the natural order is characterized by the law of karma, which governs the cause-and-effect relationships that shape individual experiences and destinies. By understanding the nature of suffering (dukkha) and the cessation of craving (nirvana), practitioners can transcend the cycle of samsara (the cycle of birth and death) and attain liberation from suffering.

Taoism, originating in ancient China, emphasizes the concept of the Tao, which can be understood as the natural order or the way of nature. The Tao Te Ching, attributed to the sage Laozi, teaches that the Tao is the underlying principle that governs all existence, beyond human comprehension. By aligning oneself with the Tao and embracing wu-wei (effortless action), individuals can harmonize with the natural order and achieve inner peace and spontaneity.

In Confucianism, the natural order is closely linked to the concept of li, which refers to ritual propriety, social norms, and ethical conduct. Confucian philosophy emphasizes the importance of maintaining social harmony and order through the cultivation of virtue (ren) and the practice of filial piety (xiao). By adhering to the principles of benevolence, righteousness, propriety, wisdom, and faithfulness, individuals contribute to the stability and prosperity of society and uphold the natural order.

Across these Eastern philosophical traditions, there is a common theme of interconnectedness, harmony, and balance within the natural order. Unlike Western notions of dominance over nature, Eastern philosophy emphasizes the importance of humility, reverence, and respect for the interconnected web of life. By aligning oneself with the rhythms and patterns of nature, individuals can cultivate inner peace, wisdom, and enlightenment, leading to a deeper understanding of the natural order and one's place within it.

Natural Order in Confucianism

Confucianism, one of the most influential philosophical and ethical systems in East Asia, offers profound insights into the concept of the natural order. Rooted in the teachings of Confucius (551-479 BCE) and later developed by his followers, Confucian philosophy emphasizes the importance of social harmony, moral cultivation, and ethical conduct. At the heart of Confucian thought lies the notion of the natural order, which is intricately connected to the principles of li, ren, and xiao.

Li, often translated as ritual propriety or ritual norms, is a central concept in Confucianism that governs social interactions, ceremonies, and relationships. Li encompasses the proper conduct and behavior expected of individuals in various social roles and contexts. It emphasizes the importance of maintaining harmony, order, and decorum in society. Through the observance of li, individuals contribute to the stability and cohesion of the community, thereby upholding the natural order.

Ren, commonly translated as benevolence or humaneness, is another key principle in Confucian philosophy. Ren emphasizes compassion, empathy, and kindness toward others. It entails treating others with respect, dignity, and understanding, regardless of their social status or background. By cultivating ren within oneself and practicing virtuous behavior, individuals contribute to the realization of social harmony and the preservation of the natural order.

Xiao, often translated as filial piety or reverence for one's parents and ancestors, is another fundamental concept in Confucian ethics. Xiao emphasizes the importance of familial relationships and the honoring of one's parents and ancestors. It entails showing gratitude, obedience, and respect toward one's parents and fulfilling one's familial duties. By upholding the virtue of xiao, individuals contribute to the stability and continuity of family life, which is seen as a microcosm of society and the natural order.

Confucianism also emphasizes the importance of education and self-cultivation in maintaining the natural order. Confucius believed in the transformative power of education in shaping individuals' moral character and social behavior. Through the pursuit of learning and self-improvement, individuals can cultivate virtue, wisdom, and moral integrity, thereby contributing to the betterment of society and the preservation of the natural order.

The natural order in Confucianism is not static but dynamic, subject to change and adaptation in response to historical, social, and cultural developments. Confucian scholars throughout history have interpreted and applied the principles of Confucianism in diverse contexts, seeking to address the changing needs and challenges of society while upholding the core values of harmony, righteousness, and benevolence.

In summary, Confucianism offers a rich and nuanced understanding of the natural order, emphasizing the importance of li, ren, xiao, education, and self-cultivation in maintaining social harmony, moral integrity, and ethical conduct. By adhering to these principles, individuals contribute to the realization of an orderly and harmonious society that is in alignment with the rhythms and patterns of the natural world.

Taoism's View on the Natural Order

Taoism, an ancient Chinese philosophical and spiritual tradition, offers a unique perspective on the concept of the natural order. Rooted in the teachings of Laozi and Zhuangzi, Taoism emphasizes living in harmony with the Tao, or the Way, which is the underlying principle that governs the universe. Central to Taoist philosophy is the notion of wu wei, or effortless action, which involves aligning oneself with the natural flow of life and allowing things to unfold spontaneously. In Taoism, the natural order is seen as a manifestation of the Tao, and individuals are encouraged to embrace simplicity, spontaneity, and non-interference in order to live in accordance with it.

The concept of the natural order in Taoism is closely linked to the idea of Yin and Yang, which represent the complementary and interconnected forces of the universe. Yin represents the passive, receptive, and feminine aspects of nature, while Yang represents the active, assertive, and masculine aspects. The balance and harmony between Yin and Yang are essential for maintaining the natural order and ensuring the smooth functioning of the cosmos. Taoist teachings emphasize the importance of embracing both Yin and Yang within oneself and in one's environment to achieve harmony and balance.

Taoism also emphasizes the importance of spontaneity and naturalness in action. Instead of striving for control or domination over nature, Taoist philosophy encourages individuals to cultivate a sense of inner stillness and receptivity, allowing them to respond to life's challenges with grace and ease. This approach to life is embodied in the principle of wu wei, which teaches that true mastery comes from non-action or effortless action. By relinquishing the need to exert excessive effort or force, individuals can align themselves with the natural rhythms of the Tao and flow with the currents of life.

Nature plays a central role in Taoist thought, serving as a source of inspiration, wisdom, and guidance. Taoist sages often retreat into nature to contemplate its beauty and observe its patterns, seeking to gain insight into the workings of the Tao. Nature is seen as a teacher and a mirror, reflecting the inherent wisdom and spontaneity of the Tao. By attuning themselves to nature's rhythms and cycles, Taoists seek to harmonize their own lives with the natural order, cultivating a deep sense of connection and reverence for the world around them.

In Taoist cosmology, the natural order is viewed as inherently fluid and ever-changing. The Tao is said to be formless and boundless, transcending all dualities and categories. As such, the natural order is not fixed or predetermined but is constantly in flux, adapting and evolving in response to the interplay of Yin and Yang and the myriad forces of existence. Taoist philosophy teaches individuals to embrace change and uncertainty, recognizing that life is a dynamic process of growth and transformation.

Overall, Taoism offers a profound and holistic understanding of the natural order, emphasizing the importance of harmony, balance, and spontaneity in living in accordance with the Tao. By embracing the principles of wu wei, Yin and Yang, and reverence for nature, individuals can cultivate a deep sense of connection with the natural world and harmonize their lives with the rhythms of the universe.

Natural Order and Theism

The concept of natural order, rooted in philosophy and metaphysics, has long been a subject of contemplation and inquiry in theism—the belief in a divine being or beings. Theistic perspectives on natural order vary widely across different religious traditions, each offering unique insights into the relationship between the natural world and the divine.

In monotheistic religions such as Christianity, Islam, and Judaism, the notion of natural order is often intricately linked to the idea of a single, all-powerful deity who created and sustains the universe. According to the Judeo-Christian tradition, for example, God is portrayed as the ultimate source of order and harmony in the cosmos. The Book of Genesis describes God's act of creation, wherein he brings order out of chaos by separating light from darkness, land from sea, and establishing the cycle of day and night. In this narrative, the natural order is seen as a reflection of God's wisdom, goodness, and providential care for his creation.

Similarly, in Islam, the Quran describes Allah as the creator of the heavens and the earth, who established the natural order with precision and purpose. Muslims believe that the universe operates according to divine laws and principles, which govern the behavior of all living beings and natural phenomena. The concept of tawhid, or the oneness of God, underscores the interconnectedness of all things within the natural order, reflecting the unity and harmony of the divine plan.

In Hinduism, a polytheistic tradition with a rich tapestry of myths and legends, the natural order is conceived in terms of cosmic cycles and eternal principles. Hindu cosmology envisions the universe as an ever-repeating cycle of creation, preservation, and dissolution, guided by the cosmic principles of dharma (righteousness), karma (action), and samsara (rebirth). The gods and goddesses of Hindu mythology are seen as manifestations of the divine within the natural world, playing various roles in upholding the cosmic order and maintaining balance and harmony.

In indigenous and animistic religions, which attribute spiritual significance to all aspects of the natural world, the concept of natural order takes on a deeply holistic and interconnected perspective. Indigenous cultures often view the Earth as a sacred and sentient being, inhabited by a multitude of spiritual beings and forces. The natural order is perceived as a delicate web of relationships between humans, animals, plants, and the elements, all of which are interconnected and interdependent. Rituals, ceremonies, and practices are often performed to honor and maintain the balance and harmony of the natural world, ensuring the well-being of both human and non-human inhabitants.

Overall, theistic perspectives on natural order offer rich insights into the relationship between the divine and the natural world. Whether conceived as the handiwork of a single creator deity, the expression of cosmic principles, or the interconnected web of life, the natural order serves as a source of wonder, awe, and contemplation for believers across diverse religious traditions. Through rituals, prayers, and spiritual practices, theistic communities seek to deepen their connection to the divine within the context of the natural world, recognizing the inherent sacredness and beauty of creation.

Natural Order in Monotheist Traditions

Natural order, a fundamental concept in philosophy and theology, holds a significant place within the monotheistic traditions of Christianity, Islam, and Judaism. Rooted in the belief in a single, omnipotent deity, these traditions offer unique perspectives on the natural world and its inherent orderliness.

In Christianity, the concept of natural order is deeply intertwined with the theological understanding of God as the creator and sustainer of the universe. According to the Christian narrative, God brought order out of chaos through the act of creation, as depicted in the Book of Genesis. The orderly progression of events, from the separation of light and darkness to the formation of land and sea, reflects God's wisdom and divine plan for the cosmos. The principle of natural order is further emphasized in the Christian doctrine of providence, which asserts that God governs the world with purpose and benevolence, guiding all things according to his will.

Similarly, in Islam, the Quran affirms the idea of natural order as a manifestation of Allah's divine wisdom and power. Muslims believe that Allah created the heavens and the earth in perfect harmony and balance, establishing precise laws and principles to govern the universe. The natural world, with its intricate ecosystems and cycles, is seen as a reflection of Allah's creative and sustaining power. The concept of tawhid, or the oneness of God, underscores the interconnectedness of all things within the natural order, emphasizing the unity and coherence of the divine plan.

In Judaism, the concept of natural order is central to the theological understanding of God's relationship with the world. The Hebrew Bible portrays God as the sovereign ruler of the universe, who established the natural laws and ordinances that govern all aspects of creation. Jewish tradition emphasizes the concept of bereshit, or divine creation, as described in the Book of Genesis. According to Jewish theology, the natural order reflects God's ongoing presence and involvement in the world, providing a framework for understanding the purpose and meaning of existence.

Across these monotheistic traditions, the concept of natural order serves as a source of awe, wonder, and reverence for the divine. It inspires believers to contemplate the beauty and complexity of the natural world, recognizing it as a reflection of God's wisdom and goodness. Moreover, the principle of natural order shapes ethical and moral teachings within these traditions, guiding individuals to live in harmony with the created order and fulfill their responsibilities as stewards of the earth.

In conclusion, natural order holds a central place within the monotheistic traditions of Christianity, Islam, and Judaism, shaping believers' understanding of God, the world, and their place within it. Rooted in the belief in a single, all-powerful deity, these traditions affirm the inherent orderliness and purposefulness of the natural world, inviting individuals to contemplate the divine wisdom and providence manifested in creation. Through rituals, prayers, and ethical teachings, monotheistic communities seek to deepen their appreciation for the natural order and cultivate a sense of reverence and stewardship towards the earth and all living beings.

Pantheism and the Natural Order

Pantheism, a philosophical and religious worldview that identifies the universe with God or equates God with the totality of existence, offers a unique perspective on the concept of natural order. Unlike monotheistic traditions that posit a separate, transcendent deity, pantheism holds that the divine and the natural world are one and the same, interconnected and inseparable.

In pantheistic thought, the natural order is not merely a reflection of divine will or design but is itself divine. The universe, with all its complexities and patterns, is seen as an expression of the inherent orderliness and harmony of existence. Pantheists often view nature as sacred and imbued with spiritual significance, believing that every aspect of the natural world possesses intrinsic value and worth.

One of the key tenets of pantheism is the idea of immanence, which asserts that the divine is present within and throughout the natural world. This perspective emphasizes the interconnectedness of all things and encourages individuals to cultivate a deep sense of reverence and respect for the environment. Pantheists often advocate for ecological stewardship and conservation, recognizing the intrinsic value of nature and the need to preserve its delicate balance.

The concept of natural order in pantheism is closely tied to notions of interconnectedness and unity. Rather than viewing the universe as a collection of separate and distinct entities, pantheists see it as a unified whole, with each part intimately connected to the others. This holistic perspective encourages individuals to recognize their interconnectedness with all living beings and to cultivate a sense of compassion and empathy towards the natural world.

Pantheism also emphasizes the cyclical nature of existence, reflecting the rhythms and patterns observed in nature. The changing seasons, the cycles of birth and death, and the endless flow of time are seen as manifestations of the natural order, illustrating the dynamic and ever-evolving nature of the universe. Pantheists often draw inspiration from these natural cycles, embracing change and impermanence as essential aspects of life.

In addition to its philosophical implications, pantheism has also influenced various cultural and artistic movements throughout history. Artists, poets, and writers have drawn inspiration from the natural world, seeking to capture its beauty and majesty in their works. Pantheistic themes can be found in literature, music, and visual art, reflecting humanity's enduring fascination with the mysteries of existence.

In conclusion, pantheism offers a profound and holistic perspective on the concept of natural order, emphasizing the interconnectedness and unity of all things. By recognizing the divine within the natural world, pantheists seek to cultivate a deep sense of reverence, awe, and respect for the environment. Through ecological stewardship, spiritual practice, and creative expression, adherents of pantheism strive to honor the sacredness of nature and live in harmony with the natural order.

Evolutionary Theories and Natural Order

Evolutionary theories have played a significant role in shaping our understanding of natural order, providing insights into the mechanisms by which life has developed and diversified over millions of years. While the concept of natural order has long been a subject of philosophical inquiry, evolutionary theory offers a scientific framework for understanding the patterns and processes that govern the natural world.

At the heart of evolutionary theory is the idea of descent with modification, first proposed by Charles Darwin in his seminal work, "On the Origin of Species." According to Darwin's theory of natural selection, organisms that are better adapted to their environment are more likely to survive and reproduce, passing on their advantageous traits to future generations. Over time, this process leads to the gradual evolution of species, resulting in the diversity of life forms observed today.

One of the key implications of evolutionary theory for our understanding of natural order is the concept of adaptation. Organisms evolve traits that enable them to thrive in specific environments, leading to a remarkable fit between organisms and their ecological niches. From the camouflage of a chameleon to the streamlined body of a dolphin, examples of adaptation abound in nature, illustrating the intricate interplay between organisms and their surroundings.

Evolutionary theory also sheds light on the interconnectedness of all living things. Through the process of descent with modification, organisms share a common ancestry, with each species branching off from a common ancestor through a series of evolutionary changes. This interconnectedness underscores the unity of life on Earth, highlighting the shared heritage that unites all living beings.

Moreover, evolutionary theory provides a framework for understanding the origins of complexity and diversity in the natural world. By elucidating the mechanisms of mutation, genetic drift, and natural selection, evolutionary biologists have unraveled the mysteries of how life has evolved from simple single-celled organisms to the vast array of species that inhabit the planet today. This understanding of the evolutionary process enriches our appreciation of the intricate web of life and the natural order that governs it.

In addition to its scientific implications, evolutionary theory has also sparked philosophical reflections on the nature of existence. The idea of natural selection as a blind and impersonal process has prompted discussions about the role of chance and contingency in shaping the course of evolution. Some philosophers have pondered the implications of evolution for our understanding of purpose and meaning in the universe, grappling with questions about the nature of existence and our place within it.

Overall, evolutionary theories have deepened our understanding of natural order, offering insights into the patterns, processes, and principles that govern life on Earth. By elucidating the mechanisms of adaptation, interconnectedness, and diversity, evolutionary theory enriches our appreciation of the complexity and beauty of the natural world. As our understanding of evolution continues to advance, so too will our appreciation of the intricate web of life and the natural order that underlies it.

The Darwinian Principle and its Philosophy

The Darwinian principle, rooted in Charles Darwin's theory of evolution by natural selection, has profoundly influenced our understanding of the natural world and its philosophical implications. At its core, the Darwinian principle posits that organisms best adapted to their environment are more likely to survive and reproduce, leading to the gradual evolution of species over time. This concept has far-reaching implications for our understanding of biology, ecology, and the nature of existence itself.

One of the key philosophical implications of the Darwinian principle is its challenge to traditional notions of teleology or purpose in nature. Unlike earlier conceptions that posited a predetermined plan or design guiding the development of life, Darwin's theory suggests that the diversity and complexity of life arise through a process of blind variation and natural selection. This perspective underscores the contingent and contingent nature of evolution, highlighting the role of chance events and environmental factors in shaping the course of life's history.

Moreover, the Darwinian principle has sparked debates about the nature of agency and causality in the natural world. While organisms appear to exhibit purposeful behavior and adaptation to their environment, the mechanisms underlying these phenomena are fundamentally non-teleological. Instead of invoking supernatural or transcendent forces, the Darwinian principle attributes the apparent design in nature to the cumulative effects of natural selection acting on heritable variation.

Furthermore, the Darwinian principle has profound implications for our understanding of human nature and society. Darwin's theory suggests that humans, like all living beings, are products of natural selection, shaped by the same evolutionary forces that have shaped the rest of life on Earth. This perspective challenges anthropocentric views of humanity and emphasizes our interconnectedness with the rest of the natural world.

In addition to its scientific and philosophical implications, the Darwinian principle has also influenced ethical and moral debates. Evolutionary biologists and philosophers have explored the evolutionary origins of human morality, seeking to understand how ethical principles and social behaviors have evolved to enhance reproductive fitness and promote cooperative interactions within groups. This perspective highlights the deep-seated biological roots of human sociality and underscores the importance of empathy, reciprocity, and cooperation in fostering a cohesive society.

Moreover, the Darwinian principle has implications for our understanding of ecology and environmental conservation. By elucidating the mechanisms of adaptation and natural selection, evolutionary theory provides insights into the dynamics of ecosystems and the importance of biodiversity for ecosystem resilience and stability. This perspective underscores the interconnectedness of all living things and the need for sustainable management of natural resources to ensure the health and well-being of future generations.

Overall, the Darwinian principle has revolutionized our understanding of the natural world and its philosophical implications. By emphasizing the role of natural selection in shaping the diversity and complexity of life, Darwin's theory has challenged traditional views of purpose and design in nature, while providing insights into the mechanisms underlying biological diversity, human nature, and social behavior. As our understanding of evolution continues to evolve, so too will its profound implications for philosophy, ethics, and our relationship with the natural world.

Natural Order and Genetics

The intersection of natural order and genetics offers profound insights into the mechanisms governing the diversity and complexity of life on Earth. Genetics, the study of heredity and variation in living organisms, provides a molecular basis for understanding how traits are passed from one generation to the next. When viewed through the lens of natural order philosophy, genetics reveals the intricate workings of evolution and the fundamental principles that govern the natural world.

At the heart of genetics lies the genetic code, a universal language encoded in DNA molecules that carries the instructions for building and operating organisms. The genetic code is composed of sequences of nucleotide bases—adenine (A), cytosine (C), guanine (G), and thymine (T)—that form the building blocks of genes. Genes are segments of DNA that contain the instructions for synthesizing proteins, the molecular machines that perform a vast array of functions within cells.

The process of inheritance ensures that genetic information is passed from parent to offspring, allowing for the transmission of traits across generations. Through mechanisms such as meiosis and fertilization, genetic variation is introduced through the shuffling and recombination of genetic material, contributing to the diversity of life. This process of genetic inheritance is central to the Darwinian principle of natural selection, which operates on the variation present within populations to drive evolutionary change over time.

Genetics also provides insights into the mechanisms of adaptation and speciation—the processes by which organisms evolve new traits and give rise to new species. Natural selection acts on genetic variation within populations, favoring individuals with traits that enhance their survival and reproductive success in a given environment. Over time, these advantageous traits become more common in the population, leading to the gradual adaptation of organisms to their ecological niches.

Furthermore, advances in molecular genetics have shed light on the molecular mechanisms underlying evolutionary processes. Research into gene expression, genetic regulation, and epigenetics has revealed the intricate networks of interactions that govern the development and functioning of organisms. These insights have deepened our understanding of how genetic variation arises and how it can lead to phenotypic diversity within and between species.

Moreover, genetics plays a crucial role in various fields of science and medicine, from agriculture and biotechnology to personalized medicine and genetic counseling. By studying the genetic basis of traits and diseases, scientists can develop novel approaches for crop improvement, disease prevention, and treatment. In agriculture, genetic engineering techniques allow for the modification of crop plants to enhance yield, nutritional content, and resistance to pests and diseases. In medicine, genetic testing enables the identification of individuals at risk for genetic disorders, allowing for early intervention and personalized treatment strategies.

In summary, the study of genetics provides a powerful framework for understanding the mechanisms of inheritance, adaptation, and evolution—the cornerstones of natural order philosophy. By unraveling the molecular basis of genetic variation and inheritance, genetics illuminates the workings of natural selection and the processes that shape the diversity of life on Earth. Moreover, genetics has practical applications in fields ranging from agriculture and medicine to conservation and biotechnology, offering solutions to some of the most pressing challenges facing humanity and the natural world.

The Intersection of Natural Order and Ethics

The intersection of natural order and ethics represents a complex and multifaceted domain where philosophical inquiries delve into the moral implications of natural phenomena and the ethical principles that guide human behavior within the natural world.

Natural order, as a philosophical concept, pertains to the inherent structure and regularity observed in the universe, suggesting that there are fundamental laws governing the behavior of natural phenomena. This concept has been a subject of contemplation for philosophers throughout history, inspiring inquiries into the nature of reality, the origins of existence, and the meaning of life. From the ancient Greeks to contemporary thinkers, the notion of natural order has served as a cornerstone of philosophical discourse, shaping perspectives on ethics and morality.

Ethics, on the other hand, is concerned with determining what is morally right or wrong and guiding human conduct based on principles of virtue, duty, and consequence. Ethical theories provide frameworks for evaluating human actions and making decisions that align with moral values and societal norms. The intersection of natural order and ethics thus involves examining the ethical implications of natural phenomena and discerning how ethical principles can be derived from or applied to the natural world.

One aspect of the intersection between natural order and ethics is the exploration of ethical considerations in ecology and environmental philosophy. Environmental ethics, a branch of philosophy concerned with the moral relationship between humans and the environment, seeks to address questions regarding the value of nature, the ethical treatment of animals, and the responsibilities of human beings towards the natural world. Philosophers such as Aldo Leopold and Rachel Carson have articulated ethical principles grounded in an understanding of natural processes and ecological systems, advocating for the preservation of biodiversity and the sustainable use of natural resources.

Moreover, discussions on natural order and ethics often intersect in debates surrounding biotechnology, genetic engineering, and other forms of scientific intervention in the natural world. Ethical dilemmas arise when considering the implications of altering or manipulating natural processes for human benefit, such as genetically modifying crops, cloning animals, or editing the human genome. Philosophers and ethicists grapple with questions of autonomy, justice, and the sanctity of life in determining the ethical boundaries of scientific innovation and technological advancement.

Furthermore, the concept of natural order has profound implications for moral philosophy and the understanding of human nature. Philosophers such as Thomas Hobbes and Jean-Jacques Rousseau have examined the relationship between human behavior and natural instincts, positing contrasting views on the inherent goodness or depravity of human beings in a state of nature. These inquiries into the natural order of human society inform ethical theories on social contract, justice, and the role of government in regulating human conduct.

In conclusion, the intersection of natural order and ethics encompasses a wide range of philosophical inquiries into the moral implications of natural phenomena, the ethical treatment of the environment, and the relationship between human beings and the natural world. By exploring the interplay between natural laws and ethical principles, philosophers seek to deepen our understanding of morality, human behavior, and the ethical responsibilities inherent in our interactions with the natural world.

Ethical Decisions in Alignment with the Natural Order

Ethical decision-making often involves navigating complex moral dilemmas and weighing the consequences of actions against a backdrop of philosophical principles and societal norms. When considering ethical decisions in alignment with the natural order, individuals and communities grapple with questions of harmony, balance, and the intrinsic value of the natural world.

At its core, the concept of the natural order suggests that there is an inherent structure and regularity in the universe, governed by fundamental laws and principles. This understanding of the natural world as an interconnected and interdependent system shapes ethical considerations in various domains, from environmental conservation to human behavior.

One aspect of ethical decisions in alignment with the natural order pertains to environmental stewardship and sustainability. Recognizing the finite resources of the planet and the delicate balance of ecosystems, individuals and organizations are increasingly adopting ethical principles that prioritize the preservation of biodiversity, the reduction of pollution, and the mitigation of climate change. This involves making choices that minimize harm to the environment, promote ecological resilience, and foster sustainable practices in industries ranging from agriculture and energy production to transportation and urban planning.

Furthermore, ethical decisions aligned with the natural order extend to considerations of animal welfare and the ethical treatment of sentient beings. Philosophical traditions such as utilitarianism, deontology, and virtue ethics provide frameworks for evaluating the moral status of animals and determining ethical responsibilities towards non-human creatures. Advocates for animal rights and welfare argue for ethical treatment based on principles of compassion, empathy, and respect for the intrinsic value of all living beings, thereby seeking to reduce suffering and promote the well-being of animals within human societies.

Moreover, ethical decisions in alignment with the natural order encompass questions of human health and well-being, particularly in relation to lifestyle choices, dietary habits, and healthcare practices. Philosophical perspectives on the connection between human health and the natural environment inform ethical considerations regarding issues such as food production, access to clean water, and the use of natural resources for medicinal purposes. This includes advocating for sustainable agriculture, promoting plant-based diets, and supporting holistic approaches to healthcare that prioritize prevention and wellness.

In addition, ethical decisions in alignment with the natural order often intersect with cultural beliefs, indigenous wisdom, and spiritual traditions that emphasize the interconnectedness of all life forms and the importance of living in harmony with nature. Indigenous cultures around the world have long held deep respect for the natural world and have developed ethical guidelines and customary practices that reflect an intimate relationship with the land, the elements, and the cycles of life.

Overall, ethical decision-making in alignment with the natural order involves a holistic approach that considers the interconnectedness of human actions with the broader web of life. By recognizing the intrinsic value of nature, promoting environmental sustainability, and fostering compassion towards all living beings, individuals and societies can strive to create a more ethical and harmonious relationship with the natural world, thereby contributing to the well-being of present and future generations.

Relativism versus Constant Ethics in Natural Order

Relativism and constant ethics represent contrasting philosophical perspectives on morality and ethical behavior within the framework of the natural order. While relativism acknowledges the diversity of cultural values and ethical norms, constant ethics posits universal moral principles that transcend cultural and historical contexts. Understanding the implications of these perspectives is essential for exploring the complexities of ethical decision-making in alignment with the natural order.

Relativism, as a philosophical stance, asserts that moral values are contingent upon cultural, societal, and individual perspectives. According to relativists, ethical judgments are subjective and context-dependent, varying across different cultures, traditions, and historical periods. This perspective recognizes the diversity of human experiences and acknowledges that what is considered morally acceptable or unacceptable in one cultural context may differ from another. Relativism highlights the importance of understanding and respecting cultural differences in ethical beliefs and practices, as well as the need for tolerance and open-mindedness in moral discourse.

In contrast, constant ethics, or moral absolutism, posits the existence of universal and unchanging moral principles that apply to all individuals, regardless of cultural or situational factors. Proponents of constant ethics argue that certain moral truths are self-evident and immutable, grounded in reason, logic, or divine command. These universal principles serve as a foundation for ethical decision-making and provide objective criteria for evaluating the rightness or wrongness of actions. Advocates of constant ethics emphasize the importance of upholding moral absolutes in the face of cultural relativism, asserting that moral truths transcend cultural diversity and historical contingencies.

When applied to the concept of the natural order, relativism and constant ethics offer distinct perspectives on how ethical decisions should be made in relation to the environment, living beings, and the broader ecosystem. Relativism may lead to a plurality of ethical perspectives, allowing for flexibility and adaptation to diverse environmental and cultural contexts. It recognizes that different cultures may have unique understandings of the natural world and ethical obligations towards it, and that these perspectives should be respected and considered in ethical deliberations.

On the other hand, constant ethics provides a framework for identifying universal moral principles that guide human interactions with the natural order. This perspective may emphasize principles such as stewardship, respect for biodiversity, and the intrinsic value of nature, which are believed to be universally applicable across cultures and contexts. Constant ethics calls for a steadfast commitment to ethical ideals, even in the face of cultural relativism or societal pressures to prioritize short-term interests over long-term sustainability.

In practice, ethical decision-making in alignment with the natural order often involves navigating the tension between relativism and constant ethics. While relativism encourages openness to diverse perspectives and cultural values, constant ethics emphasizes the importance of upholding moral principles that transcend cultural relativism. Finding a balance between these two perspectives requires critical reflection, empathy, and a willingness to engage in dialogue with diverse stakeholders to promote ethical behavior and environmental sustainability. Ultimately, both relativism and constant ethics offer valuable insights into the complexities of morality within the natural order, highlighting the need for nuanced and context-sensitive approaches to ethical decision-making in today's interconnected world.

Natural Order as Depicted in Literature

The concept of the natural order has long been a recurring theme in literature, serving as a source of inspiration, contemplation, and philosophical inquiry for writers across different cultures and historical periods. In literature, the natural order often symbolizes the harmonious balance and inherent laws governing the universe, as well as the interconnectedness between humanity, nature, and the cosmos.

Throughout literary history, various works have depicted the natural order in diverse ways, reflecting cultural beliefs, philosophical ideologies, and artistic interpretations of the relationship between humans and the natural world. One notable example is William Shakespeare's play "King Lear," where the disruption of the natural order is a central theme. As Lear's kingdom descends into chaos and disorder, symbolized by natural phenomena such as storms and earthquakes, the protagonist's journey becomes a metaphor for the consequences of human folly and the need to restore balance to the natural order.

Similarly, the Romantic poets of the 19th century, including William Wordsworth and Samuel Taylor Coleridge, often explored the theme of the natural order in their works. In poems such as Wordsworth's "Lines Composed a Few Miles above Tintern Abbey" and Coleridge's "Kubla Khan," nature is depicted as a sublime and transcendent force that inspires awe and reverence in the human soul. These poets celebrated the beauty and majesty of the natural world while reflecting on humanity's place within it, suggesting a deep connection between the individual psyche and the cosmic order.

In modern literature, the concept of the natural order continues to be a prominent motif, appearing in works ranging from classic novels to contemporary fiction. In J.R.R. Tolkien's "The Lord of the Rings," for example, the natural order is reflected in the balance of power between different races and species, as well as the interconnectedness of the land of Middle-earth with its inhabitants. Tolkien's intricate world-building and rich mythology highlight the importance of preserving the natural order in the face of external threats and internal conflicts.

Another example can be found in contemporary eco-literature, which explores themes of environmental degradation, ecological awareness, and humanity's responsibility towards the natural world. Authors such as Barbara Kingsolver, Margaret Atwood, and Cormac McCarthy have addressed these issues in their novels, using the natural order as a backdrop to examine the ethical, social, and political implications of human interaction with the environment.

Furthermore, in fantasy literature, authors often draw upon the concept of the natural order to create elaborate fictional worlds inhabited by mythical creatures, magical beings, and supernatural phenomena. In J.K. Rowling's "Harry Potter" series, for instance, the wizarding world operates according to its own set of natural laws and magical principles, which coexist alongside the mundane reality of the Muggle world. Rowling's depiction of the natural order in her novels underscores the importance of balance, harmony, and respect for the natural world, even in a fantastical setting.

Overall, the portrayal of the natural order in literature reflects humanity's enduring fascination with the mysteries of the universe and our ongoing quest to understand our place within it. Whether through poetry, prose, or fantasy, writers continue to explore the complexities of the natural order, inviting readers to contemplate the beauty, wonder, and interconnectedness of all living things in the grand tapestry of existence.

Shakespeare's Contribution to the Understanding of Natural Order

William Shakespeare, widely regarded as one of the greatest playwrights in history, made significant contributions to the understanding of the natural order through his works. Across his plays, Shakespeare explores the themes of order and disorder, harmony and chaos, and the delicate balance of the universe. Through his characters and their experiences, Shakespeare offers profound insights into human nature and the cosmic order.

In many of Shakespeare's plays, the disruption of the natural order serves as a central plot device, driving the narrative forward and shaping the characters' fates. One of the most notable examples is found in "King Lear," where the titular character's decision to divide his kingdom among his daughters leads to a series of tragic events that ultimately culminate in chaos and destruction. As Lear's kingdom descends into disorder, symbolized by raging storms and political upheaval, the play explores the consequences of violating the natural order and the need for restoration and reconciliation.

Similarly, in "Macbeth," Shakespeare delves into the theme of ambition and its consequences on the natural order. The protagonist, Macbeth, driven by his desire for power and glory, disrupts the divine order by committing regicide and usurping the throne. As a result, the natural world is thrown into turmoil, with unnatural occurrences such as supernatural visitations and disturbances in the animal kingdom serving as omens of impending doom. Through Macbeth's tragic downfall, Shakespeare highlights the importance of moral integrity and the dangers of succumbing to unchecked ambition.

In "Hamlet," Shakespeare explores the theme of revenge and its impact on the natural order. The protagonist, Hamlet, is torn between his duty to avenge his father's murder and his moral qualms about committing regicide. As the play unfolds, Hamlet grapples with questions of justice, morality, and the inherent disorder of the human condition. Through Hamlet's existential crisis, Shakespeare probes the complexities of human nature and the fragile balance between order and chaos.

Moreover, Shakespeare's comedies also engage with the concept of the natural order, albeit in a lighter and more whimsical manner. In plays such as "A Midsummer Night's Dream" and "As You Like It," Shakespeare explores themes of love, marriage, and social hierarchy within the framework of the natural world. Through mistaken identities, romantic entanglements, and comic misunderstandings, Shakespeare celebrates the inherent harmony of nature and the joyous resolution of conflict.

Overall, Shakespeare's contribution to the understanding of the natural order lies in his nuanced exploration of human behavior, moral dilemmas, and the eternal struggle between order and disorder. Through his timeless works, Shakespeare invites readers and audiences to reflect on the complexities of existence and the enduring relevance of the natural order in shaping our lives and destinies.

Modern Literary Takes on the Natural Order

In modern literature, authors continue to explore the concept of the natural order, drawing inspiration from philosophical traditions while infusing their works with contemporary insights and perspectives. Through their storytelling, these writers offer unique interpretations of the natural order and its implications for human existence, society, and the world at large.

One notable example of modern literary takes on the natural order is found in the works of Cormac McCarthy, particularly his novel "The Road." Set in a post-apocalyptic world ravaged by an unspecified disaster, "The Road" follows the journey of a father and son as they navigate a desolate landscape in search of safety and salvation. Against the backdrop of environmental devastation and societal collapse, McCarthy explores themes of survival, morality, and the precarious balance of life and death. Through the stark realism of his prose, McCarthy highlights the fragility of the natural order and the resilience of the human spirit in the face of adversity.

Another contemporary author who grapples with the concept of the natural order is Margaret Atwood, known for her dystopian novel "The Handmaid's Tale." Set in a totalitarian society where women are subjugated and reproductive rights are severely restricted, "The Handmaid's Tale" explores themes of power, control, and the erosion of fundamental freedoms. Atwood's depiction of a world governed by rigid social hierarchies and oppressive norms serves as a cautionary tale about the dangers of disrupting the natural order and the consequences of unchecked authority.

Furthermore, the works of Haruki Murakami offer a unique perspective on the natural order, blending elements of magical realism with existential themes. In novels such as "Kafka on the Shore" and "1Q84," Murakami blurs the boundaries between reality and fantasy, exploring the interconnectedness of human consciousness and the mysteries of the universe. Through his surreal narratives and enigmatic characters, Murakami invites readers to contemplate the nature of existence and the inherent order that underlies the chaos of everyday life.

Additionally, contemporary authors like Zadie Smith and Salman Rushdie infuse their works with rich cultural tapestries and diverse perspectives on the natural order. In novels such as "White Teeth" and "Midnight's Children," Smith and Rushdie explore themes of identity, heritage, and the clash of cultures in a globalized world. Through their intricate storytelling and vibrant characters, these authors shed light on the complexities of human relationships and the fluidity of social constructs within the natural order.

Overall, modern literary takes on the natural order reflect the diverse range of human experiences and perspectives in an ever-changing world. From dystopian visions of societal collapse to surreal explorations of existential mysteries, contemporary authors continue to grapple with the timeless themes of order, chaos, and the human condition. Through their thought-provoking works, these writers offer readers new insights into the intricate web of life and the enduring quest for meaning and understanding.

Natural Order in Political Philosophy

Natural order in political philosophy refers to the concept of an inherent and immutable structure or arrangement that governs the organization and functioning of society. Throughout history, political philosophers have grappled with the idea of natural order, seeking to understand its implications for governance, justice, and human behavior.

One of the earliest discussions of natural order in political philosophy can be found in the works of ancient Greek thinkers such as Plato and Aristotle. Plato, in his dialogue "The Republic," proposed the idea of a natural hierarchy in society, with philosopher-kings ruling over the masses based on their innate wisdom and virtue. Aristotle, meanwhile, explored the concept of natural law, arguing that there are universal principles that govern human conduct and political organization.

During the Enlightenment period, political philosophers such as John Locke and Jean-Jacques Rousseau further developed ideas related to natural order. Locke, in his "Two Treatises of Government," posited that individuals are born with certain natural rights, including the right to life, liberty, and property. According to Locke, government exists to protect these rights, and any system that violates them is illegitimate. Rousseau, on the other hand, emphasized the importance of the general will in determining the legitimate authority of government. He argued that true democracy arises when individuals collectively consent to be governed by the general will, which reflects the common interests of society.

In the 19th century, the concept of natural order underwent further scrutiny and debate. Karl Marx, in his critique of capitalism, challenged traditional notions of natural hierarchy and argued for the abolition of class distinctions. According to Marx, the capitalist system disrupts the natural order by exploiting the labor of the working class for the benefit of the ruling bourgeoisie. He envisioned a society in which the means of production are collectively owned and controlled, leading to a more equitable distribution of wealth and power.

In more contemporary political philosophy, scholars continue to explore the concept of natural order in relation to issues such as social justice, human rights, and environmental sustainability. Theories of distributive justice, for example, seek to establish principles for allocating resources and opportunities in a fair and equitable manner. Environmental ethics, meanwhile, grapple with the question of how humans should relate to the natural world and whether there are inherent values or obligations that guide our interactions with the environment.

Moreover, discussions surrounding globalization, multiculturalism, and technological advancement have prompted reevaluations of traditional conceptions of natural order. As societies become increasingly interconnected and diverse, questions arise about the universality of moral principles and the compatibility of different cultural norms and values. Similarly, advances in science and technology raise ethical dilemmas regarding issues such as genetic engineering, artificial intelligence, and biotechnology, challenging our understanding of what it means to be human and how we should govern ourselves in an ever-changing world.

In conclusion, the concept of natural order in political philosophy is a complex and multifaceted idea that has evolved over time in response to changing social, political, and intellectual contexts. From ancient Greek theories of natural hierarchy to contemporary debates about social justice and environmental ethics, political philosophers continue to explore the implications of natural order for governance, justice, and human flourishing.

Political Systems and the Natural Order

Political systems and the concept of natural order have long been intertwined, as societies seek to establish governance structures that reflect their understanding of inherent principles and values. Throughout history, various political systems have emerged, each influenced by different interpretations of the natural order and its implications for social organization and governance.

One of the earliest political systems to grapple with the idea of natural order was feudalism, which dominated much of medieval Europe. Feudalism was characterized by a hierarchical structure in which power and authority were decentralized, with kings, lords, and vassals each holding specific rights and responsibilities. The feudal system was often justified by appeals to divine or natural order, with monarchs claiming their authority was ordained by God or determined by hereditary lineage.

During the Renaissance and Enlightenment periods, the rise of nation-states led to the development of new political systems based on principles of sovereignty, citizenship, and popular sovereignty. Absolutist monarchies, such as those in France and Russia, asserted the divine right of kings to rule, while constitutional monarchies, such as those in England and the Netherlands, established limits on royal authority and granted rights to citizens through written constitutions and parliamentary institutions.

The Enlightenment also saw the emergence of republican forms of government, inspired by classical models of democracy from ancient Greece and Rome. Philosophers such as Montesquieu and Rousseau advocated for systems of government based on the consent of the governed and the separation of powers, in which legislative, executive, and judicial functions were divided to prevent tyranny and ensure accountability.

In the modern era, the spread of democracy and liberalism has led to the establishment of democratic political systems in many parts of the world. Democracy, derived from the Greek words "demos" (people) and "kratos" (rule), emphasizes the importance of popular sovereignty and majority rule in decision-making processes. Liberal democracies, such as those found in Western Europe and North America, also prioritize individual rights and freedoms, including freedom of speech, assembly, and religion, as essential components of the natural order.

However, not all political systems adhere to principles of democracy and liberalism. Authoritarian regimes, such as those in China, North Korea, and Saudi Arabia, prioritize stability and order over individual rights and freedoms. These systems often justify their authority through appeals to traditional values, national unity, or divine mandate, rather than the consent of the governed.

In recent years, debates about the natural order and political systems have intensified with the rise of globalization, technological advancement, and social change. Issues such as income inequality, climate change, and social justice have prompted reevaluations of existing political systems and calls for reform. Some advocate for greater government intervention to address societal challenges, while others argue for decentralization and deregulation to promote individual freedom and economic growth.

In conclusion, political systems and the concept of natural order are closely intertwined, as societies seek to establish governance structures that reflect their understanding of inherent principles and values. From feudalism to democracy, political systems have evolved over time in response to changing social, economic, and cultural contexts, shaping the way societies are organized and governed.

Anarchism and the Philosophy of Natural Order

Anarchism, often misunderstood and misrepresented, is a political philosophy that challenges traditional notions of governance and authority. Rooted in the belief in individual freedom and voluntary cooperation, anarchism offers a unique perspective on the concept of natural order, advocating for a society based on mutual aid, solidarity, and decentralized decision-making.

At its core, anarchism rejects the idea of a centralized state or hierarchical authority, viewing such institutions as oppressive and detrimental to human freedom. Instead, anarchists propose a society organized around principles of direct democracy, community self-governance, and voluntary association. In this vision, individuals are free to pursue their interests and aspirations without interference from coercive institutions or structures of power.

The philosophy of natural order plays a central role in anarchist thought, providing a foundation for understanding how societies can function without authoritarian control. Anarchists argue that human beings have an innate capacity for cooperation and mutual aid, which can form the basis of social organization in the absence of state authority. They point to examples of spontaneous order and self-organization in nature, such as ecosystems and social insects, as evidence that complex systems can emerge without top-down control.

Anarchism encompasses a diverse range of perspectives and approaches, including individualist anarchism, social anarchism, and anarcha-feminism, each offering its own interpretation of how natural order can be achieved in human societies. Individualist anarchists, such as Max Stirner and Benjamin Tucker, emphasize the importance of personal autonomy and voluntary exchange, advocating for a society based on individual sovereignty and free association.

Social anarchists, on the other hand, focus on collective action and solidarity, seeking to abolish hierarchical structures of power and create a society based on principles of equality and mutual aid. Figures like Mikhail Bakunin, Peter Kropotkin, and Emma Goldman are prominent voices within social anarchism, advocating for forms of organization such as workers' councils, communes, and federations as alternatives to the state.

Anarcha-feminism, a branch of anarchism that emerged in the late 19th and early 20th centuries, explores the intersections between patriarchy, capitalism, and state power, arguing that the struggle for gender liberation is inseparable from the fight against hierarchy and domination. Anarcha-feminists like Voltairine de Cleyre and Emma Goldman advocate for the dismantling of gender norms and the creation of non-hierarchical relationships based on mutual respect and autonomy.

Critics of anarchism often raise concerns about the viability of anarchist societies in the face of social complexity and potential conflicts. However, anarchists argue that decentralized decision-making, voluntary association, and direct democracy can address these challenges more effectively than centralized authority. By empowering individuals and communities to govern their own affairs, anarchism seeks to create a society that respects the diversity of human experience and fosters cooperation and solidarity.

In conclusion, anarchism offers a compelling vision of natural order based on principles of individual freedom, voluntary cooperation, and mutual aid. Rooted in the belief that humans are capable of organizing themselves without hierarchical authority, anarchism challenges traditional notions of governance and offers alternative models of social organization. While anarchism faces criticism and skepticism, its emphasis on decentralization, autonomy, and solidarity continues to inspire movements for social change and liberation around the world.

Natural Order and Environmental Ethics

Natural order and environmental ethics are intimately connected concepts that shape our understanding of the relationship between humanity and the natural world. While natural order refers to the inherent organization and harmony found in nature, environmental ethics explores the moral obligations and responsibilities that humans have towards the environment.

At its core, natural order encompasses the intricate balance and interdependence of all living organisms and ecosystems on Earth. It recognizes the interconnectedness of life and the delicate equilibrium that sustains biodiversity and ecological stability. From the microscopic interactions between soil microbes to the complex relationships within entire ecosystems, natural order reflects the inherent harmony and orderliness of the natural world.

Environmental ethics, on the other hand, is concerned with the moral principles and values that guide human interactions with the environment. It seeks to address questions of environmental justice, sustainability, and stewardship, considering the ethical implications of human actions on the natural world. Environmental ethics draws from various philosophical traditions, including utilitarianism, deontology, and virtue ethics, to develop frameworks for understanding our moral responsibilities towards nature.

One of the key principles of environmental ethics is the recognition of the intrinsic value of nature, independent of its utility to humans. This perspective challenges anthropocentric views that prioritize human interests over the well-being of other species and ecosystems. Instead, it emphasizes the inherent worth of all living beings and ecosystems, advocating for their protection and preservation for their own sake.

Natural order provides a philosophical foundation for environmental ethics, highlighting the importance of maintaining ecological balance and harmony in human interactions with the environment. It underscores the interconnectedness of all life forms and the importance of preserving biodiversity and ecosystem integrity. From this perspective, environmental degradation and habitat destruction disrupt the natural order, leading to ecological imbalances and potential harm to both humans and non-human beings.

Environmental ethics also considers the concept of sustainability, recognizing the finite nature of Earth's resources and the need to manage them responsibly for future generations. This involves adopting practices that minimize environmental impact, promote ecological resilience, and ensure the long-term well-being of ecosystems and communities. Sustainability aligns with the principles of natural order by emphasizing the importance of maintaining ecological balance and harmony in human activities.

In contemporary society, issues such as climate change, deforestation, pollution, and loss of biodiversity highlight the urgent need for ethical considerations in environmental decision-making. Environmental ethics calls for a reevaluation of human values and priorities, urging individuals and societies to adopt more sustainable and environmentally responsible behaviors.

Moreover, environmental ethics encourages collective action and global cooperation to address environmental challenges on a systemic level. It emphasizes the interconnectedness of environmental issues and the need for collaborative efforts to achieve meaningful solutions. By promoting environmental awareness, fostering stewardship ethics, and advocating for policy changes, environmental ethics seeks to protect and restore the natural order for the benefit of present and future generations.

In conclusion, natural order and environmental ethics offer valuable insights into humanity's relationship with the natural world. They highlight the interconnectedness of all life forms and the importance of preserving ecological balance and harmony. By integrating ethical considerations into environmental decision-making, we can work towards a more sustainable and equitable future for both humans and the environment.

Environmental Balance and Natural Order

Environmental balance and natural order are intricately intertwined concepts that play a crucial role in shaping the health and sustainability of ecosystems worldwide. While environmental balance refers to the state of equilibrium and harmony within ecosystems, natural order encompasses the inherent organization and interdependence of all living beings and their environment.

In the natural world, ecosystems are finely tuned systems where each organism and component contributes to the overall balance and functioning of the system. From the smallest microorganisms to the largest predators, every species plays a vital role in maintaining ecological stability and resilience. This balance is achieved through complex interactions such as predation, competition, and mutualism, which regulate population sizes and resource availability.

Biodiversity is a key indicator of environmental balance, as diverse ecosystems tend to be more resilient to disturbances and better able to withstand environmental changes. High levels of biodiversity provide a buffer against ecological disruptions and enhance ecosystem productivity and stability. Conversely, disruptions to environmental balance, such as habitat destruction, pollution, and climate change, can lead to biodiversity loss and ecological degradation.

Human activities have a profound impact on environmental balance and natural order, often disrupting delicate ecosystems and causing widespread environmental degradation. Deforestation, for example, disrupts the balance of carbon dioxide and oxygen in the atmosphere, leading to climate change and loss of habitat for countless species. Similarly, pollution from industrial activities contaminates air, water, and soil, threatening the health and well-being of ecosystems and human communities alike.

Addressing environmental imbalances requires a multifaceted approach that considers the complex interactions between human societies and the natural world. Conservation efforts play a critical role in restoring and maintaining environmental balance by protecting habitats, preserving biodiversity, and promoting sustainable land and resource management practices. Conservation initiatives aim to mitigate the impacts of human activities on the environment and promote the recovery of degraded ecosystems.

Natural order provides a philosophical framework for understanding the underlying principles that govern ecosystems and guide conservation efforts. It emphasizes the interconnectedness of all living beings and the importance of preserving ecological balance and harmony. By recognizing the intrinsic value of nature and its components, natural order encourages stewardship ethics and responsible environmental decision-making.

Efforts to restore environmental balance often involve ecological restoration projects that aim to rehabilitate degraded ecosystems and enhance their resilience to future disturbances. These projects may involve reforestation, habitat restoration, and the reintroduction of native species to their natural habitats. By restoring ecological functions and processes, these initiatives help to recreate the natural order within ecosystems and promote their long-term health and sustainability.

In addition to conservation and restoration efforts, sustainable development practices are essential for achieving environmental balance and preserving natural order. Sustainable development seeks to meet the needs of present generations without compromising the ability of future generations to meet their own needs. It involves integrating economic, social, and environmental considerations into decision-making processes to ensure that development is environmentally responsible and socially equitable.

In conclusion, environmental balance and natural order are fundamental principles that guide the functioning and resilience of ecosystems worldwide. By recognizing the interconnectedness of all living beings and their environment, and by promoting stewardship ethics and sustainable development practices, we can work towards restoring and maintaining environmental balance for the benefit of present and future generations.

Human Impacts on the Natural Order

Human activities have significantly altered the natural order of ecosystems worldwide, leading to widespread environmental degradation and loss of biodiversity. From deforestation to pollution, human impacts on the natural world have far-reaching consequences that threaten the health and sustainability of ecosystems and the well-being of all living beings.

One of the most significant human impacts on the natural order is habitat destruction and fragmentation. Deforestation, urbanization, and agricultural expansion have resulted in the loss and degradation of vital habitats for countless species. As natural habitats disappear, species populations decline, and ecological balance is disrupted, leading to cascading effects throughout entire ecosystems.

Pollution is another major human impact on the natural order, affecting air, water, and soil quality. Industrial activities, transportation, and agricultural practices release pollutants into the environment, contaminating ecosystems and threatening the health of both wildlife and human populations. Pollution can disrupt ecological processes, harm aquatic life, and contribute to the decline of biodiversity.

Climate change, driven primarily by human-induced greenhouse gas emissions, poses a significant threat to the natural order of ecosystems worldwide. Rising temperatures, shifting precipitation patterns, and more frequent extreme weather events are altering habitats and disrupting ecological processes. Species are struggling to adapt to rapidly changing conditions, leading to shifts in species distributions and loss of biodiversity.

Overexploitation of natural resources is another human impact on the natural order that threatens the health and sustainability of ecosystems. Unsustainable fishing practices, poaching, and illegal wildlife trade have led to the depletion of fish stocks, decline of iconic species, and disruption of food webs. Overharvesting of forests, minerals, and other resources further exacerbates environmental degradation and undermines the natural order of ecosystems.

Invasive species represent yet another human impact on the natural order, disrupting native ecosystems and outcompeting native species for resources. Invasive plants, animals, and pathogens can spread rapidly and outcompete native species, leading to declines in biodiversity and altering ecosystem dynamics. Invasive species can also have significant economic and ecological impacts, costing billions of dollars in control efforts and damage each year.

Human impacts on the natural order are not limited to terrestrial ecosystems; marine environments are also experiencing significant degradation due to human activities. Overfishing, habitat destruction, pollution, and climate change threaten the health of marine ecosystems and the biodiversity they support. Coral reefs, mangroves, and seagrass beds are particularly vulnerable to human impacts, with profound consequences for marine life and coastal communities.

Addressing human impacts on the natural order requires collective action and a commitment to sustainable development and conservation. Conservation efforts, such as habitat restoration, protected area management, and species conservation programs, are essential for preserving biodiversity and restoring ecological balance. Sustainable resource management practices, renewable energy development, and emissions reduction initiatives are also critical for mitigating the impacts of climate change and pollution.

Ultimately, recognizing the interconnectedness of human societies and the natural world is essential for promoting environmental stewardship and preserving the natural order for future generations. By adopting sustainable lifestyles, supporting conservation efforts, and advocating for policies that prioritize environmental protection, we can work towards restoring and maintaining the delicate balance of ecosystems worldwide.

Natural Order and Quantum Physics

The concept of natural order, rooted in philosophy and metaphysics, has long been a subject of inquiry into the fundamental principles governing the universe. In recent decades, the exploration of natural order has extended into the realm of quantum physics, where scientists grapple with the peculiar behaviors of particles at the subatomic level.

Quantum physics, a branch of physics that deals with the behavior of matter and energy on the smallest scales, challenges conventional notions of causality and determinism that underpin classical physics. At the heart of quantum mechanics is the principle of uncertainty, famously encapsulated in Heisenberg's uncertainty principle, which states that the more precisely one property of a particle is measured, the less precisely another complementary property can be known.

This inherent uncertainty in the behavior of quantum particles raises profound questions about the nature of reality and the concept of natural order. In classical physics, events are assumed to unfold according to deterministic laws, where the present state of a system uniquely determines its future evolution. However, quantum mechanics introduces an element of randomness into the fabric of reality, challenging our understanding of cause and effect and the notion of a predetermined natural order.

One of the key features of quantum physics that challenges traditional notions of natural order is the phenomenon of superposition. According to quantum mechanics, particles such as electrons can exist in multiple states simultaneously until measured, at which point their wavefunction collapses into a single state. This idea suggests that particles can exist in a state of indeterminacy, defying the classical idea of a fixed and deterministic natural order.

Another intriguing aspect of quantum physics is entanglement, a phenomenon in which the states of two or more particles become correlated in such a way that the state of one particle instantaneously influences the state of another, regardless of the distance between them. This non-local connection between entangled particles violates the principle of locality and challenges our intuitive notions of causality and the natural order of events.

Furthermore, quantum physics introduces the concept of wave-particle duality, which suggests that particles such as photons and electrons can exhibit both particle-like and wave-like behavior depending on the experimental setup. This duality blurs the distinction between particles and waves and underscores the fundamental uncertainty inherent in the quantum world.

While quantum mechanics may seem at odds with our classical understanding of natural order, some physicists argue that it offers a more profound and nuanced perspective on the nature of reality. Rather than rejecting the concept of natural order outright, quantum physics challenges us to reevaluate our assumptions and adopt a more flexible and probabilistic view of the universe.

In conclusion, the exploration of natural order in the context of quantum physics reveals the complex and mysterious nature of reality at the smallest scales. While quantum mechanics introduces uncertainty and randomness into our understanding of the universe, it also offers profound insights into the interconnectedness of all things and the fundamental principles that govern the cosmos. As scientists continue to probe the mysteries of quantum physics, our conception of natural order may evolve to encompass a broader and more inclusive understanding of the underlying principles that shape our universe.

Reality and the Perception of Natural Order

Reality is a multifaceted concept that encompasses the physical world, subjective experiences, and the interplay between perception and interpretation. At the heart of our understanding of reality lies the notion of natural order, a fundamental principle that governs the structure and behavior of the universe. However, the perception of natural order is deeply influenced by individual perspectives, cultural beliefs, and cognitive biases, leading to diverse interpretations of reality.

From a philosophical standpoint, reality is often conceived as an objective and immutable framework that exists independently of human observation. According to this view, natural order represents the underlying structure of the universe, characterized by fundamental laws and principles that govern the behavior of matter and energy. These laws, such as the laws of physics and chemistry, provide a framework for understanding the dynamics of the natural world and the relationships between different phenomena.

However, the perception of natural order is not limited to objective observation but is also shaped by subjective experiences and cognitive processes. Human perception is inherently selective and interpretive, filtering sensory information through individual beliefs, emotions, and cognitive biases. As a result, individuals may perceive and interpret the same reality in different ways, leading to subjective variations in the perception of natural order.

Cultural and societal influences further shape the perception of natural order, as cultural beliefs and traditions often dictate how individuals interpret and make sense of the world around them. Different cultures may have distinct cosmological frameworks, creation myths, and philosophical traditions that inform their understanding of reality and natural order. These cultural perspectives contribute to the diversity of human experience and the richness of interpretations of natural order.

Furthermore, advances in psychology and cognitive science have shed light on the role of perception and cognition in shaping our understanding of reality. The concept of perceptual constancy, for example, highlights how our brains interpret sensory information to maintain a stable and coherent perception of the world, despite changes in environmental conditions. This process of perceptual interpretation influences our perception of natural order and contributes to the construction of our subjective reality.

In addition to individual and cultural influences, technological advancements have also impacted the perception of natural order. The advent of virtual reality, for instance, allows individuals to immerse themselves in simulated environments that may deviate from physical reality. Virtual worlds challenge our conventional understanding of natural order by blurring the boundaries between physical and virtual experiences, raising questions about the nature of reality and the limits of human perception.

Overall, the perception of natural order is a complex and multifaceted phenomenon that is influenced by a variety of factors, including individual experiences, cultural beliefs, cognitive processes, and technological advancements. While the objective reality may exist independently of human observation, our subjective interpretation of natural order is inherently shaped by our perceptions, biases, and cultural context. Understanding the interplay between perception and natural order is essential for gaining insight into the diversity of human experience and the complex nature of reality.

The Role of Uncertainty in Natural Order

Uncertainty plays a significant role in our understanding of natural order, challenging traditional notions of stability and predictability in the universe. While natural order implies a sense of regularity and consistency, the presence of uncertainty introduces complexity and dynamism into the fabric of reality. From the microscopic realm of quantum mechanics to the macroscopic scale of cosmology, uncertainty manifests in various forms, shaping our perception of the natural world and the philosophical concepts associated with it.

At the heart of uncertainty in natural order lies the principle of indeterminacy, a fundamental aspect of quantum mechanics first articulated by Werner Heisenberg in his famous uncertainty principle. According to this principle, certain pairs of physical properties, such as position and momentum, cannot be simultaneously measured with arbitrary precision. Instead, there exists inherent uncertainty in the precise values of these properties, leading to probabilistic descriptions of physical phenomena. This indeterminacy challenges the classical notion of determinism and underscores the inherent unpredictability of quantum systems.

In addition to quantum uncertainty, chaos theory explores the behavior of dynamical systems that are highly sensitive to initial conditions, leading to unpredictable and seemingly random outcomes. Chaos theory reveals how deterministic systems can exhibit behavior that appears random due to their complex nonlinear dynamics. The concept of the "butterfly effect," popularized by Edward Lorenz, illustrates how small changes in initial conditions can lead to drastically different outcomes over time, highlighting the role of sensitivity to initial conditions in shaping the evolution of natural systems.

Uncertainty also permeates our understanding of cosmology, where the nature of the universe's origin, composition, and fate remains subject to ongoing scientific inquiry and debate. The cosmic microwave background radiation, relic radiation from the early universe, provides valuable insights into the universe's past and its subsequent evolution. However, uncertainties in cosmological parameters, such as the density of dark matter and dark energy, introduce uncertainty into our understanding of cosmic structure and dynamics.

Philosophically, uncertainty challenges traditional notions of determinism and necessitates a more nuanced understanding of natural order. While deterministic models seek to predict future states of systems based on their initial conditions and governing laws, uncertainty introduces inherent limitations to our predictive capabilities. This recognition of uncertainty underscores the complexity and richness of natural systems, highlighting the need for probabilistic and stochastic frameworks to describe their behavior accurately.

Moreover, uncertainty fosters humility and open-mindedness in our approach to understanding the natural world, recognizing the inherent limitations of human knowledge and perception. As scientific inquiry continues to push the boundaries of our understanding, uncertainty serves as a catalyst for discovery and innovation, driving us to explore new frontiers and challenge existing paradigms.

In conclusion, uncertainty is an integral aspect of natural order, pervading all levels of scientific inquiry and philosophical reflection. From the microscopic realm of quantum mechanics to the macroscopic scale of cosmology, uncertainty challenges our understanding of determinism and introduces complexity into the fabric of reality. Embracing uncertainty allows us to appreciate the richness and diversity of natural systems, fostering humility and curiosity in our exploration of the mysteries of the universe.

Contemporary Criticisms of the Natural Order Concept

Critiques of the concept of natural order have emerged in contemporary philosophical discourse, challenging traditional assumptions about the inherent harmony and balance of the natural world. While the idea of a natural order suggests a fundamental structure or pattern underlying reality, critics argue that this notion is fraught with philosophical, ethical, and scientific complexities that warrant closer examination.

One key criticism of the concept of natural order revolves around its potential to perpetuate hierarchical and oppressive social structures. Critics argue that appeals to a supposed natural order have been historically used to justify inequalities based on race, gender, and socioeconomic status. For example, the notion of a "natural hierarchy" has been invoked to justify discriminatory practices and policies, reinforcing existing power dynamics and marginalizing certain groups within society. In this view, the concept of natural order serves as a rhetorical tool to legitimize social inequality rather than promote justice and equity.

Moreover, contemporary criticisms of the natural order concept often intersect with concerns about environmental degradation and ecological sustainability. Critics argue that traditional conceptions of natural order tend to prioritize human interests over the well-being of the broader ecosystem, leading to unsustainable exploitation of natural resources and disruption of ecological balance. This anthropocentric view of nature neglects the interconnectedness of all living beings and fails to account for the long-term consequences of human actions on the environment.

In addition to ethical concerns, critics also question the scientific validity of the natural order concept, particularly in light of advancements in complexity theory and systems science. The complex, nonlinear dynamics of natural systems defy simplistic notions of order and predictability, challenging the idea of a predetermined or static natural order. Instead, contemporary science recognizes the inherent unpredictability and dynamism of natural processes, emphasizing the need for probabilistic and adaptive models to describe complex phenomena accurately.

Furthermore, critiques of the natural order concept often highlight its potential to stifle innovation and progress by imposing rigid constraints on human behavior and social organization. By prescribing certain norms or values as inherently "natural" or "immutable," proponents of natural order risk inhibiting social change and diversity of thought. Instead, critics advocate for a more fluid and inclusive understanding of nature that accommodates diverse perspectives and values human agency in shaping the future trajectory of society and the environment.

In response to these criticisms, scholars and philosophers have proposed alternative frameworks for understanding the relationship between humanity and the natural world. One such approach is ecological philosophy, which emphasizes the interconnectedness and interdependence of all living beings within ecosystems. By recognizing the intrinsic value of nature and promoting ecological stewardship, ecological philosophy seeks to foster a more harmonious and sustainable relationship between humans and the environment.

Overall, contemporary criticisms of the natural order concept highlight the need for a more nuanced and context-sensitive understanding of nature and society. By acknowledging the complexities and limitations of traditional conceptions of natural order, we can engage in more meaningful dialogue about how to address pressing social, ethical, and environmental challenges in the modern world.

The Naturalistic Fallacy

The naturalistic fallacy is a concept in philosophy that warns against the logical error of deriving moral judgments or ethical norms from purely descriptive statements about the natural world. In essence, it is the mistaken belief that what is "natural" is inherently good or right, while what is "unnatural" is inherently bad or wrong. This fallacy is closely related to discussions about natural order and has been a subject of debate among ethicists and philosophers for centuries.

The naturalistic fallacy was first articulated by the British philosopher G.E. Moore in his seminal work "Principia Ethica" published in 1903. Moore argued that it is fallacious to equate the concept of "goodness" with any natural property or attribute because the two are fundamentally distinct. He famously illustrated this point with his open-question argument, which invites us to consider whether the naturalistic claim "X is good because it is natural" can be sensibly challenged by asking "Is X really good?" Despite the numerous attempts to define "good" in terms of natural properties, Moore asserted that such definitions inevitably fail to capture the full meaning of moral goodness.

The naturalistic fallacy poses significant challenges for ethical theories that seek to ground morality in facts about the natural world, such as evolutionary ethics or utilitarianism. Critics argue that these theories commit the naturalistic fallacy by conflating descriptive claims about human behavior or biological traits with normative claims about what is morally right or wrong. For example, proponents of evolutionary ethics may argue that certain moral behaviors, such as altruism or cooperation, are "natural" because they confer evolutionary advantages. However, critics contend that such arguments fail to bridge the gap between descriptive and normative claims, thereby succumbing to the naturalistic fallacy.

Moreover, the naturalistic fallacy has important implications for contemporary debates in bioethics, environmental ethics, and public policy. For instance, proponents of certain biomedical interventions or genetic modifications may appeal to their naturalness as a justification for their ethical permissibility. However, critics caution against relying solely on naturalness as a criterion for ethical evaluation, emphasizing the need to consider broader ethical principles and consequences.

In environmental ethics, the naturalistic fallacy underscores the complexity of defining what is "natural" in the context of human interventions in the environment. While some argue that human activities should align with natural processes and ecosystems, others caution against romanticizing the concept of nature or privileging it over human well-being. Instead, environmental ethicists advocate for a more nuanced understanding of humanity's relationship with the natural world, one that balances ecological integrity with human needs and values.

In conclusion, the naturalistic fallacy serves as a reminder of the inherent limitations of deriving moral judgments from factual statements about the natural world. By recognizing the distinction between descriptive and normative claims, ethicists and philosophers can engage in more rigorous and meaningful discussions about the foundations of morality and ethics. Ultimately, a thoughtful and reflective approach to ethical reasoning requires careful consideration of a wide range of factors beyond mere appeals to nature.

Postmodernist Views

Postmodernism is a philosophical and cultural movement that emerged in the mid-20th century and profoundly influenced various fields, including philosophy, literature, art, architecture, and social sciences. Unlike traditional philosophical perspectives, postmodernism challenges the idea of fixed truths, objective reality, and universal principles, including those related to natural order.

One of the central tenets of postmodernist thought is the rejection of grand narratives or overarching explanations of the world. Instead, postmodernists argue that reality is fragmented, subjective, and contingent upon individual perspectives and experiences. In the context of natural order, postmodernists question the existence of a universally applicable order or structure inherent in the natural world. They argue that attempts to impose such order are often rooted in power dynamics and social constructs rather than objective truth.

Postmodernist views on natural order often intersect with critiques of modernity and Enlightenment rationalism. Critics of modernity argue that the Enlightenment's emphasis on reason, progress, and scientific objectivity has led to the domination of nature and the marginalization of diverse ways of knowing and being. Postmodernists challenge the idea of a fixed and hierarchical natural order, highlighting the plurality of human experiences and the multiplicity of perspectives on the world.

In literature and art, postmodernism is characterized by experimentation, intertextuality, and the blurring of boundaries between genres, forms, and disciplines. Postmodernist writers and artists often employ techniques such as pastiche, irony, and fragmentation to reflect the fragmented nature of reality and to subvert traditional narratives of order and meaning. In this context, natural order is depicted as a construct shaped by language, culture, and power relations rather than an objective reality.

Postmodernist philosophers, such as Jacques Derrida and Michel Foucault, have critiqued the notion of natural order from different perspectives. Derrida's deconstructionist approach challenges the stability of meaning and the binary oppositions that underpin traditional conceptions of order. According to Derrida, language is inherently unstable, and attempts to establish fixed meanings ultimately give rise to hierarchies and exclusionary practices.

Foucault's work on power and knowledge further complicates the idea of natural order by examining how discourses and institutions shape our understanding of the world.

Foucault's concept of the "order of things" highlights the historical contingency of knowledge and the ways in which power operates through systems of classification and categorization.

In architecture and urban planning, postmodernism has led to the rejection of modernist notions of order and uniformity in favor of diversity, plurality, and hybridity. Postmodern architects often incorporate eclectic styles, historical references, and playful elements into their designs, challenging the notion of a singular natural order in the built environment.

Overall, postmodernist views on natural order offer a critical perspective on the assumptions and ideologies that underpin traditional conceptions of reality. By emphasizing plurality, contingency, and multiplicity, postmodernism encourages us to question dominant narratives and to recognize the diversity of human experiences and perspectives. In this way, postmodernism invites us to reconsider our understanding of natural order and to embrace the complexities and contradictions inherent in the world around us.

Natural Order and Artificial Intelligence

Natural order, a concept deeply rooted in philosophical discourse, pertains to the inherent structure and organization believed to exist within the natural world. As humanity progresses into the digital age, the intersection between natural order and artificial intelligence (AI) presents intriguing questions and challenges. AI, a branch of computer science, involves the development of systems capable of performing tasks that typically require human intelligence, including learning, problem-solving, and decision-making. Exploring the relationship between natural order and AI unveils a complex interplay between human ingenuity and the fundamental principles governing the universe.

At its core, natural order encompasses the principles of causality, regularity, and predictability observed in the natural world. These principles form the basis of scientific inquiry and provide a framework for understanding and interpreting phenomena. In contrast, AI operates within the realm of human-created systems, which may not necessarily adhere to the same principles as the natural world. While AI algorithms strive to mimic aspects of human cognition and behavior, they do not inherently possess an understanding of natural order. Instead, AI systems rely on data inputs, algorithms, and computational power to perform tasks, often diverging from the intricacies of natural phenomena.

However, the development of AI has sparked discussions about its potential to uncover insights into the natural order. Machine learning algorithms, a subset of AI, have demonstrated the ability to analyze vast amounts of data and identify patterns that may elude human observation. By processing data from diverse sources, AI systems can uncover correlations, trends, and relationships that contribute to our understanding of the underlying structure of the natural world. For example, AI-powered simulations have been used to model complex systems such as weather patterns, ecological dynamics, and molecular interactions, shedding light on the interconnectedness of natural phenomena.

Moreover, AI technologies have the potential to enhance our ability to study and preserve the natural environment. Remote sensing technologies, equipped with AI algorithms, enable scientists to monitor ecosystems, track changes in biodiversity, and detect environmental hazards with unprecedented accuracy and efficiency. By harnessing AI capabilities, researchers can gather valuable insights into the dynamics of ecosystems, contributing to conservation efforts and sustainable management practices.

Despite these advancements, ethical considerations surrounding AI and natural order persist. The increasing reliance on AI raises concerns about the potential consequences of delegating decision-making processes to autonomous systems. While AI algorithms excel at processing data and identifying patterns, they may lack the ethical judgment and moral reasoning inherent in human decision-making. Moreover, biases inherent in data sets used to train AI models can perpetuate social inequalities and reinforce existing power dynamics, leading to unintended consequences for individuals and communities.

Furthermore, the pursuit of AI-driven solutions must be accompanied by a commitment to ethical and responsible development practices. As AI technologies continue to evolve, stakeholders must prioritize transparency, accountability, and human-centered design principles to ensure that AI systems align with societal values and respect human dignity. By integrating ethical considerations into the design, deployment, and governance of AI technologies, we can mitigate potential risks and harness the transformative potential of AI to advance our understanding of the natural order while upholding ethical standards and promoting the well-being of humanity and the planet.

In summary, the relationship between natural order and artificial intelligence reflects a dynamic interplay between human innovation and the fundamental principles governing the universe. While AI technologies offer unprecedented opportunities to explore and understand the complexities of the natural world, they also present ethical challenges that must be addressed. By fostering interdisciplinary collaboration and ethical reflection, we can leverage AI to deepen our appreciation of the natural order while safeguarding the values and principles that define humanity's relationship with the world.

Technology and the Challenge of Natural Order

In the intricate tapestry of human history, technology has emerged as a potent force reshaping our understanding and interaction with the natural world. The concept of natural order, deeply embedded in philosophical discourse, refers to the inherent structure and organization believed to govern the universe. However, the relentless march of technological advancement presents a profound challenge to this perceived order, raising questions about humanity's role in shaping the fabric of existence.

Throughout the ages, humans have sought to harness the power of technology to transcend the limitations imposed by the natural world. From the invention of the wheel to the development of sophisticated artificial intelligence, technological innovations have enabled humanity to conquer new frontiers and redefine the boundaries of possibility. However, this quest for mastery over nature has often come at a cost, as human activities disrupt the delicate equilibrium of ecosystems and alter the course of natural processes.

One of the most pressing challenges posed by technology is its impact on the environment. Industrialization, fueled by technological advancements, has led to widespread pollution, deforestation, and habitat destruction, threatening biodiversity and destabilizing ecosystems. The relentless extraction of natural resources to fuel technological progress has strained the planet's finite resources, leading to concerns about sustainability and the long-term viability of human civilization.

Moreover, the proliferation of digital technologies has given rise to a new set of challenges related to information overload, privacy concerns, and the erosion of human connection. The advent of social media, for example, has transformed the way we communicate and interact, blurring the lines between the virtual and physical worlds. While technology has the potential to enhance communication and foster collaboration, it also poses risks such as misinformation, cyberbullying, and addiction.

Furthermore, the rapid pace of technological innovation has raised ethical questions about the implications of artificial intelligence, genetic engineering, and other emerging technologies. As humans gain the ability to manipulate the building blocks of life and create intelligent machines, we must grapple with profound questions about the nature of consciousness, free will, and the boundaries of human agency. The development of autonomous weapons, for instance, raises concerns about the ethical implications of delegating life-and-death decisions to machines.

Despite these challenges, technology also holds the promise of addressing pressing global issues such as climate change, poverty, and disease. Renewable energy technologies offer a pathway to a more sustainable future, while advances in healthcare have the potential to improve quality of life and extend human lifespan. Furthermore, digital technologies can empower marginalized communities, enhance education, and foster economic development, contributing to a more equitable and inclusive society.

In navigating the complex relationship between technology and the natural order, it is essential to adopt a holistic and multidisciplinary approach. By integrating insights from philosophy, ethics, science, and engineering, we can develop technologies that are not only innovative and efficient but also sustainable and aligned with the principles of natural order. Moreover, fostering a culture of responsible innovation and ethical reflection can help ensure that technology serves the collective good and promotes the well-being of both humanity and the planet.

In conclusion, the challenge of technology and its impact on the natural order is one of the defining issues of our time. As we stand at the crossroads of technological progress and environmental stewardship, it is imperative that we approach these challenges with humility, foresight, and a deep respect for the intricate web of life that sustains us. By harnessing the transformative power of technology while honoring the principles of natural order, we can forge a more harmonious relationship with the natural world and build a brighter future for generations to come.

Ethical Questions in the Age of AI

In the rapidly evolving landscape of artificial intelligence (AI), ethical questions loom large, challenging our understanding of human values, autonomy, and responsibility. As AI technologies become increasingly integrated into our daily lives, from virtual assistants to autonomous vehicles, society grapples with complex moral dilemmas that demand thoughtful consideration and informed decision-making.

One of the foremost ethical concerns surrounding AI revolves around issues of accountability and transparency. As AI systems make decisions that affect individuals and communities, questions arise about who bears responsibility for their actions. Unlike human decision-makers, AI algorithms lack consciousness and moral agency, raising concerns about how to assign accountability in cases of algorithmic bias, errors, or unintended consequences. Moreover, the opaque nature of many AI algorithms makes it difficult to understand how decisions are made and to detect and address instances of bias or discrimination.

Another pressing ethical question pertains to the impact of AI on employment and the economy. While AI has the potential to boost productivity, efficiency, and innovation, it also poses risks such as job displacement, income inequality, and economic disruption. As automation replaces human labor in various industries, policymakers and businesses must grapple with the ethical implications of technological unemployment and the need for reskilling and retraining programs to ensure a smooth transition to a more automated future.

Furthermore, the ethical use of AI raises concerns about privacy, surveillance, and data protection. AI systems often rely on vast amounts of personal data to train and improve their algorithms, raising questions about consent, data ownership, and the right to privacy. Moreover, the use of AI-powered surveillance technologies, such as facial recognition systems, raises concerns about the erosion of civil liberties and the potential for abuse by authoritarian regimes or unscrupulous actors.

In addition to these concerns, AI also presents ethical challenges related to fairness, justice, and equity. AI algorithms can perpetuate and exacerbate existing biases and inequalities, leading to discriminatory outcomes in areas such as hiring, lending, and criminal justice. Moreover, the lack of diversity in the tech industry and the underrepresentation of marginalized communities in AI development exacerbate these issues, highlighting the need for greater inclusivity and diversity in the design and deployment of AI systems.

Another ethical dilemma arises from the potential misuse of AI for malicious purposes, such as cyberattacks, disinformation campaigns, and autonomous weapons systems. The rapid advancement of AI technology has led to concerns about its potential to destabilize global security and undermine democratic institutions. Policymakers and international organizations are grappling with the need to establish norms, regulations, and safeguards to mitigate the risks associated with the proliferation of AI-powered weapons and cyber warfare tactics.

Despite these ethical challenges, AI also holds the promise of addressing some of humanity's most pressing problems, from healthcare and education to climate change and social justice. AI-driven innovations such as personalized medicine, predictive analytics, and smart infrastructure have the potential to improve quality of life, enhance decision-making, and foster economic development. Moreover, AI can augment human capabilities, enabling us to tackle complex problems more effectively and efficiently.

In conclusion, the ethical questions surrounding AI are multifaceted and complex, touching on issues of accountability, transparency, privacy, fairness, and security. As AI continues to reshape our world, it is essential that we approach these challenges with humility, foresight, and a commitment to upholding human values and dignity. By fostering a culture of ethical reflection, responsible innovation, and inclusive governance, we can harness the transformative power of AI for the benefit of all humanity while mitigating its potential risks and pitfalls.

Reconceptualizing Natural Order in the 21st Century

In the 21st century, the concept of natural order undergoes a profound reconceptualization as society grapples with rapid technological advancements, environmental challenges, and shifting cultural paradigms. Once considered a static and immutable framework governing the cosmos, the notion of natural order now evolves into a dynamic and multifaceted concept that reflects the complexity and interconnectedness of modern life.

One of the central aspects of reconceptualizing natural order involves embracing a holistic and interdisciplinary approach that integrates insights from philosophy, science, technology, and spirituality. Instead of viewing natural order as a rigid hierarchy imposed from above, contemporary thinkers recognize it as an emergent property of complex adaptive systems, encompassing everything from the laws of physics and ecological dynamics to human behavior and societal norms.

Moreover, the reconceptualization of natural order emphasizes the interdependence and interconnectivity of all living beings and ecosystems, highlighting the need for ecological stewardship and sustainable development. In an era of unprecedented environmental degradation and climate change, the preservation of natural order requires a concerted effort to protect biodiversity, mitigate pollution, and promote ecological resilience.

Furthermore, the advent of digital technologies and artificial intelligence challenges traditional conceptions of natural order by blurring the boundaries between the organic and the synthetic. As humans increasingly manipulate and engineer living organisms and ecosystems, questions arise about the ethical implications of playing "creator" and the long-term consequences of tampering with the fabric of life.

Additionally, the reconceptualization of natural order prompts a reevaluation of human values, aspirations, and priorities in light of our evolving understanding of the cosmos and our place within it. Instead of viewing ourselves as separate from nature, we recognize our interconnectedness with all living beings and our responsibility to act as stewards of the Earth.

Furthermore, the reconceptualization of natural order necessitates a shift in our economic and political systems toward greater inclusivity, equity, and justice. In an era marked by growing inequality and social unrest, the pursuit of natural order requires us to address systemic injustices and create more sustainable and equitable societies that honor the dignity and rights of all individuals.

Moreover, the reconceptualization of natural order invites us to explore new forms of spirituality and consciousness that transcend traditional religious frameworks and embrace the interconnectedness of all life. From mindfulness practices and meditation to psychedelic therapies and transpersonal experiences, people are increasingly seeking spiritual experiences that foster a deeper connection to nature and the cosmos.

In conclusion, the reconceptualization of natural order in the 21st century represents a paradigm shift in our understanding of the cosmos, our place within it, and our responsibility to steward its resources wisely. By embracing a holistic and interdisciplinary approach, fostering ecological stewardship, promoting social justice, and cultivating a deeper spiritual connection to the natural world, we can honor the inherent beauty and complexity of life and ensure a more harmonious and sustainable future for generations to come.

Sustainability and the Natural Order

In the context of natural order philosophy, sustainability emerges as a foundational principle that guides human actions and interactions with the environment. Sustainability, at its core, embodies the concept of living in harmony with the natural world, recognizing the interconnectedness of all living beings and ecosystems, and ensuring that the needs of present generations are met without compromising the ability of future generations to meet their own needs.

At the heart of sustainability lies the recognition that the Earth operates according to intricate systems and cycles that maintain balance and order. These natural systems, from the water cycle and carbon cycle to the intricate web of biodiversity, exemplify the principles of natural order, illustrating how life on Earth is sustained through dynamic interactions and feedback loops.

Moreover, sustainability encompasses not only environmental conservation but also social equity and economic prosperity. A truly sustainable society seeks to address systemic inequalities, uplift marginalized communities, and promote economic systems that prioritize the well-being of both people and the planet. In this way, sustainability becomes a holistic endeavor that integrates environmental, social, and economic considerations.

One of the key challenges in achieving sustainability is the need to reconcile human activities with the regenerative capacity of the Earth's ecosystems. Human civilization has long relied on the exploitation of natural resources for economic growth and technological advancement, often at the expense of environmental degradation and biodiversity loss. However, as our understanding of the interconnectedness of natural systems deepens, there is a growing recognition of the need to adopt more sustainable practices that respect the limits of the Earth's carrying capacity.

In recent years, there has been a global movement towards sustainable development, driven by concerns over climate change, resource depletion, and environmental degradation. Governments, businesses, and civil society organizations are increasingly embracing sustainable practices, from renewable energy and circular economy initiatives to conservation efforts and sustainable agriculture.

Furthermore, sustainability requires a shift in mindset away from short-term thinking and immediate gratification towards long-term planning and foresight. This shift entails reevaluating our consumption patterns, investing in renewable energy and green technologies, and prioritizing conservation and environmental restoration efforts.

Education also plays a crucial role in fostering a culture of sustainability, empowering individuals with the knowledge, skills, and values needed to make informed decisions and take collective action towards a more sustainable future. By promoting environmental literacy, fostering ecological awareness, and instilling a sense of responsibility towards future generations, education can be a powerful tool for driving positive change.

Ultimately, sustainability and the natural order are deeply intertwined, reflecting humanity's ethical imperative to live in harmony with the Earth and respect the inherent interconnectedness of all life. As we navigate the complex challenges of the 21st century, embracing sustainability as a guiding principle can help us build a more resilient, equitable, and flourishing world for ourselves and future generations.

Prospects for the Future of Natural Order

The concept of natural order has been a central theme in philosophy for centuries, shaping our understanding of the world and our place within it. As we look to the future, the prospects for the continuation and evolution of natural order philosophy are both intriguing and complex.

One of the key challenges facing the future of natural order philosophy is the rapid pace of technological advancement and its impact on our relationship with the natural world. Technological innovations have led to profound changes in the way we interact with our environment, from the development of industrial agriculture and the widespread use of fossil fuels to the rise of artificial intelligence and genetic engineering. While these advancements offer the potential for greater efficiency and convenience, they also pose significant risks to the delicate balance of natural systems.

In the face of these challenges, there is a growing recognition of the need to integrate technology with principles of sustainability and ecological resilience. This involves harnessing the power of innovation to develop more sustainable solutions to pressing environmental problems, such as climate change, pollution, and biodiversity loss. From renewable energy technologies and green infrastructure to sustainable agriculture and biomimicry, there are countless opportunities to leverage technology in service of the natural order.

Moreover, as our understanding of complex systems and nonlinear dynamics continues to deepen, we are gaining new insights into the intricacies of natural order. From the study of ecological networks and ecosystem services to the emerging field of complex systems science, researchers are uncovering the underlying principles that govern the behavior of natural systems. This knowledge not only enhances our appreciation for the beauty and complexity of the natural world but also informs our efforts to manage and preserve it for future generations.

Another important aspect of the future of natural order philosophy is its intersection with ethics and morality. As we confront global challenges such as climate change, resource depletion, and social inequality, there is a growing need for ethical frameworks that can guide our actions in alignment with the principles of natural order. This involves reevaluating our values and priorities, fostering a deeper sense of interconnectedness and empathy, and embracing a more holistic approach to decision-making that takes into account the well-being of both people and the planet.

Furthermore, the future of natural order philosophy is closely intertwined with broader social and political movements that seek to promote environmental stewardship and social justice. From grassroots activism and community organizing to international agreements and policy reforms, there is a growing movement towards a more sustainable and equitable world. By mobilizing collective action and advocating for systemic change, individuals and communities can play a vital role in shaping the future of natural order philosophy.

In conclusion, the prospects for the future of natural order philosophy are both promising and challenging. While technological advancements and societal changes present new opportunities and complexities, there is a growing recognition of the importance of embracing principles of sustainability, resilience, and interconnectedness in shaping our relationship with the natural world. By integrating these principles into our ethics, policies, and practices, we can work towards a more harmonious and sustainable future for all.

The Future of Natural Order Philosophy

As humanity progresses into the future, the concept of natural order philosophy faces both challenges and opportunities. Natural order philosophy, rooted in the idea of an inherent balance and harmony in the universe, has long served as a guiding principle for understanding our place in the world and our relationship with nature. Looking ahead, several key factors will shape the future trajectory of natural order philosophy.

One of the most pressing issues facing natural order philosophy is the ongoing environmental crisis. Climate change, habitat destruction, pollution, and resource depletion threaten the delicate balance of ecosystems around the globe. In response, there is a growing recognition of the need to prioritize sustainability and conservation efforts to protect biodiversity and safeguard the natural order. This shift towards environmental stewardship is reflected in international agreements such as the Paris Agreement and in grassroots movements advocating for environmental justice and conservation.

Furthermore, advances in science and technology present both opportunities and challenges for natural order philosophy. On one hand, scientific discoveries offer new insights into the complexity and interconnectedness of natural systems, deepening our understanding of the principles underlying the natural order. Fields such as ecology, systems theory, and complexity science provide valuable frameworks for studying and managing ecosystems. On the other hand, technological advancements such as genetic engineering, artificial intelligence, and geoengineering raise ethical questions about their potential impacts on the natural world and human society.

Another important consideration for the future of natural order philosophy is the intersection with ethics and morality. As we confront global challenges such as climate change, social inequality, and technological disruption, there is a growing need for ethical frameworks that can guide our actions in alignment with the principles of the natural order. This involves reevaluating our values and priorities, fostering a deeper sense of interconnectedness and empathy, and embracing a more holistic approach to decision-making that considers the well-being of both people and the planet.

Moreover, the future of natural order philosophy is closely tied to broader social and political movements advocating for systemic change. From indigenous rights movements and environmental activism to calls for social justice and equity, there is a growing recognition of the interconnectedness of social, environmental, and economic issues. By addressing root causes of inequality and injustice and promoting sustainable and regenerative practices, these movements contribute to the realization of a more harmonious and just society in line with the principles of the natural order.

In conclusion, the future of natural order philosophy hinges on our ability to address pressing environmental challenges, integrate scientific advancements with ethical considerations, and promote social and political change towards a more sustainable and equitable world. By embracing principles of interconnectedness, balance, and harmony, we can work towards a future where humanity lives in harmony with nature and respects the inherent wisdom of the natural order.

The Role of the Natural Order in Future Ethics

As humanity advances into the future, the concept of the natural order continues to play a crucial role in shaping ethical considerations and guiding decision-making processes. The natural order philosophy, deeply rooted in the idea of harmony, balance, and interconnectedness in the universe, serves as a moral compass for navigating complex ethical dilemmas and fostering a more sustainable and equitable society.

One of the key ways in which the natural order informs future ethics is through its emphasis on ecological stewardship and sustainability. With mounting environmental challenges such as climate change, habitat destruction, and species extinction, there is a growing recognition of the need to prioritize the well-being of the planet and its ecosystems. By aligning ethical principles with the natural order, individuals and societies can work towards practices that promote the health and resilience of the environment, ensuring the long-term viability of life on Earth.

Furthermore, the natural order philosophy informs ethical considerations in the realm of social justice and human rights. Central to the concept of the natural order is the idea of interconnectedness and mutual dependence among all living beings. This perspective underscores the importance of empathy, compassion, and solidarity in addressing issues such as poverty, inequality, and discrimination. By recognizing the inherent dignity and worth of every individual and striving for equitable distribution of resources and opportunities, societies can uphold ethical principles that are in harmony with the natural order.

In addition, the natural order philosophy guides ethical decision-making in the context of technological advancement and scientific innovation. As humanity continues to develop new technologies and harness the power of science to improve our lives, it is essential to consider the potential impacts on the natural world and future generations. By adopting an ethical framework grounded in the principles of the natural order, individuals and organizations can mitigate harm, promote responsible innovation, and ensure that technological advancements are aligned with the well-being of both people and the planet.

Moreover, the natural order philosophy offers insights into ethical considerations related to cultural diversity and intercultural dialogue. Recognizing the diversity of cultural traditions and worldviews, the natural order encourages a spirit of openness, respect, and collaboration among different communities. By embracing cultural diversity and engaging in meaningful dialogue, societies can cultivate ethical principles that reflect the richness and complexity of human experience while honoring the interconnectedness of all life forms.

Finally, the natural order philosophy invites reflection on the ethical implications of our collective actions and choices as a species. As we confront global challenges such as overconsumption, environmental degradation, and social unrest, it is imperative to consider the long-term consequences of our behavior and prioritize solutions that are in alignment with the natural order. By fostering a deep sense of responsibility and stewardship towards the Earth and its inhabitants, humanity can forge a path towards a more ethical and sustainable future.

In conclusion, the natural order philosophy serves as a guiding principle for future ethics, informing ethical considerations in environmental sustainability, social justice, technological innovation, cultural diversity, and collective responsibility. By integrating the wisdom of the natural order into ethical decision-making processes, humanity can work towards creating a more harmonious, equitable, and flourishing world for present and future generations.

Natural Order in the Age of Climate Change

In the face of escalating climate change, the concept of the natural order takes on renewed significance, offering valuable insights into humanity's relationship with the environment and guiding efforts to mitigate and adapt to environmental challenges. The natural order philosophy, grounded in the interconnectedness and interdependence of all living beings, underscores the urgent need for collective action to address the impacts of climate change and safeguard the health and integrity of the planet.

At its core, the natural order philosophy emphasizes harmony and balance within the ecosystem, recognizing the delicate interplay of ecological processes that sustain life on Earth. However, human activities such as burning fossil fuels, deforestation, and industrial agriculture have disrupted this delicate balance, leading to widespread environmental degradation and accelerating climate change. Rising global temperatures, extreme weather events, melting ice caps, and rising sea levels are just some of the consequences of our unsustainable practices, threatening ecosystems, biodiversity, and the well-being of communities worldwide.

In light of these challenges, the natural order philosophy calls for a fundamental reorientation of human society towards more sustainable and regenerative practices. This entails transitioning to renewable energy sources, reducing carbon emissions, conserving natural habitats, and promoting biodiversity conservation. By aligning our actions with the principles of the natural order, we can restore ecological balance, mitigate the impacts of climate change, and protect the planet for future generations.

Furthermore, the natural order philosophy emphasizes the interconnectedness of environmental and social systems, highlighting the disproportionate impacts of climate change on vulnerable communities, including indigenous peoples, marginalized populations, and low-income communities. Inequities in access to resources, economic opportunities, and political power exacerbate the effects of climate change, perpetuating cycles of poverty and social injustice. Addressing climate change requires not only technical solutions but also social and economic reforms that promote equity, justice, and human dignity.

Moreover, the natural order philosophy encourages a holistic approach to climate change that considers the interconnectedness of environmental, social, and economic systems. This includes adopting policies and practices that prioritize the well-being of both people and the planet, fostering resilience, and promoting sustainable development. By embracing the principles of the natural order, we can build more resilient and adaptive communities capable of thriving in a changing climate.

Additionally, the natural order philosophy emphasizes the importance of humility, reverence, and gratitude towards the natural world. In the face of climate change, it calls for a shift in consciousness towards a deeper appreciation of nature's inherent value and the interconnected web of life. Cultivating a sense of kinship and stewardship towards the Earth can inspire individuals and communities to take meaningful action to protect and preserve the environment.

In conclusion, the natural order philosophy offers valuable insights and guidance in the age of climate change, reminding us of our interconnectedness with the natural world and the imperative to live in harmony with the Earth. By aligning our actions with the principles of the natural order, we can address the root causes of climate change, promote environmental sustainability, and create a more just, equitable, and resilient world for present and future generations.

Have Questions / Comments?

This book was designed to cover as much as possible but I know I have probably missed something, or some new amazing discovery that has just come out.

If you notice something missing or have a question that I failed to answer, please get in touch and let me know. If I can, I will email you an answer and also update the book so others can also benefit from it.

Thanks For Being Awesome :)

Submit Your Questions / Comments At:

https://xspurts.com/posts/questions

Get Another Book Free

We love writing and have produced a huge number of books.

For being one of our amazing readers, we would love to offer you another book we have created, 100% free.

To claim this limited time special offer, simply go to the site below and enter your name and email address.

You will then receive one of my great books, direct to your email account, 100% free!

https://xspurts.com/posts/free-book-offer

www.ingramcontent.com/pod-product-compliance
Lightning Source LLC
Chambersburg PA
CBHW050327230526
45471CB00005B/2386